Placer Mining

A HAND-BOOK FOR KLONDIKE AND OTHER MINERS AND PROSPECTORS

WITH

INTRODUCTORY CHAPTERS REGARDING THE RECENT GOLD DISCOVERIES IN THE YUKON VALLEY, THE ROUTES TO THE GOLD FIELDS, OUTFIT REQUIRED, AND MINING REGULATIONS OF ALASKA AND THE CANADIAN YUKON.

ALSO

A Map of the Yukon Valley, Embracing all the Information Obtainable from Reliable Sources up to December 1st, 1897.

SCRANTON, PA.
THE COLLIERY ENGINEER COMPANY
1897

Facsimile Reproduction 1965
The Shorey Book Store
815 Third Avenue
Seattle, Washington 98104
SJS 106

SIXTH EDITION
Of Facsimile Reprint

Limited to 500 Copies

MARCH 1974

ISBN # 0-8466-0106-0

PREFACE.

The demand for this book is the excuse for its appearance. Thousands have rushed off to the Klondike, not only ignorant of the first principles of placer mining, the only occupation at which any one can make a living in that region, but also almost entirely ignorant of the nature of the country into which they went. It is said that not more than 5 out of 100 of those who started for the Klondike in the fall of 1897 had any better information as to how to reach the diggings or what to do after they reached them than to "follow the crowd." It is needless to say that those who expected to find the journey to Dawson City a pleasure trip, or who threw away their outfits in the scramble to be first at the mines, or who expected to dig gold from the ground as the farmer digs potatoes, have long since found their mistake. Fortunately, several thousands of those who started for the mines were obliged to turn back and wait till spring. From these and from thousands of others who expect to be in the rush to Alaska in the spring and summer of 1898 has come a demand for such information as this book contains. It is believed that in this little volume is summarized all the most important and reliable information obtainable regarding the Yukon gold fields, or which can be obtained before the summer of 1898; that it contains the best map of these regions yet published, and the most practical, and, hence, the most valuable treatise on placer and hydraulic mining in print. We recommend this book to the mining public, confident that it will make friends wherever it goes.

THE COLLIERY ENGINEER CO.,
Scranton, Pa.

December, 1897.

CONTENTS.

PLACER MINING.

CHAPTER I.

THE YUKON GOLD FIELDS.

In July, 1897, one of the greatest and richest gold fields ever opened was brought to the notice of the civilized world. In this case, it did not require months of time and labor to verify the stories of wealth brought over by a hundred miners from the Klondike district, situated just across the Alaskan border in the Northwest Territory. These miners had ample evidence that gold existed in plenty where they came from, for the steamers Excelsior and Portland, on which they came, brought down $1,100,000, or more than $10,000 for each miner. This was just the beginning. During August, September, and October, steamer after steamer brought down men with sacks or valises full of the precious metal, until $2,500,000 in gold has been brought out and put into circulation. Within a year after gold was discovered on the Klondike in August, 1896, $5,500,000 had been taken out on Eldorado and Bonanza Creeks. Over half of the output was kept at the mines for use in business operations, gold dust being the only circulating medium in use there. At least $1,000,000 more would have been brought out in the fall had not the Yukon River become lower than ever before known in August, preventing the river boats from making their final trips to the diggings. As the gold is too heavy to bring out overland, it was kept at the diggings until spring.

For richness of the ground the Klondike has seldom been equaled in the history of gold mining. Single claims produced $150,000 and $200,000 during the winter of 1896, and in the spring the owners declared they had worked only small corners of their mines. Single pans of dirt (2 shovelfuls) yielded $800 and $1,000, and pans containing $300 to $500 were not uncommon.

The most authentic news obtainable places the prospective output of gold between December, 1897, and July, 1898,

DAWSON CITY.

at $20,000,000. It is predicted by conservative men who returned from the Klondike in October, 1897, that the steamers which reach Puget Sound next July will bring down $15,000,000 to $17,000,000, or about 40 tons of gold dust and nuggets.

Dr. Dawson, of the Canadian Government, after whom

Dawson City, the center of the recent gold discoveries, is named, states that he considers the Yukon destined to be the greatest mining country the world ever saw. He anticipates that the recent discoveries will lead to the development of quartz mining, in which is the staple wealth of any mining country. Experienced prospectors have already made a number of valuable gold quartz discoveries in the Yukon district, and many more are sure to follow.

Inspector Strickland, of the Yukon mounted police, in speaking of the Klondike discoveries, says:

"There has been no exaggeration. I have seen nothing in newspapers in regard to the richness of the field that is not true. Great strikes have been made, but the amount of gold is unlimited. There are hundreds of creeks rich in gold-bearing placers, never yet entered by prospectors. Of course, all the claims in the creeks now opened are taken up, but those are only beginnings, I believe, of much greater finds."

The most conservative and accurate statement of what has been accomplished on the Klondike is that made by William Ogilvie, the official Surveyor for the Dominion of Canada, who returned in September, 1897, from a two years' stay on the Yukon, this being his second trip to that country. Mr. Ogilvie has had unusual facilities for observation, having surveyed the claims on the Klondike. He says:

"When we consider the unseasonable weather, the unfavorable conditions for mining, and the still more unfavorable conditions regarding food, utensils, and labor, what has been accomplished on the Klondike may, without hazard, be asserted to be unique in the history of mining. I will do no more than say generally that we have a region which will yet be the scene of numerous mining enterprises, both placer and quartz, the latter practically inexhaustible.

This country under more favorable conditions would be the richest and most extensive mining area in the world to-day. Notwithstanding the disadvantages of long, cold winters and lack of roads, we have here a wide field for profitable investment and room for thousands of happy, contented, cultivated homes."

OTHER YUKON DISTRICTS.

The Klondike is but one of several valuable placer deposits in the Yukon Valley, among which are the Hootalinqua, Stewart, MacMillan, Forty-Mile, Sixty-Mile, Birch Creek, Munook Creek, Tanana, and Koyukuk districts.

These rich gold fields extend along the Yukon Valley for upwards of 1,200 miles, and lie on both sides of the boundary between Alaska and Canada. Some of the richest placer deposits in the world lie in the neighborhood of Circle City, Alaska, and were abandoned in the rush to the Klondike last winter. Valuable placer deposits have recently been found on Munook Creek, 450 miles west of Circle City, and the recent discovery of gold mines in Siberia shows that they are in the same belt. These mines are in the same chain of mountains that supplied California's gold, and in the same general line with the Peru fields. Running up the coast, this gold vein traverses Alaska and, crossing Bering Strait, crops out again in Siberia. The whole Alaskan region is, therefore, within the gold belt, and there is scarcely a place within its confines where gold has been diligently sought in which it has not been found in greater or less quantity. American miners and prospectors are advised to expend their energies in discovering the deposits which exist on the American side of the line, as they will then be free from any exactions and restrictions imposed by the Canadian Government.

In the summer of 1896, the United States Government had a geological survey expert make a thorough investigation of the Alaskan gold fields. The result of the exploration is stated by the director of the survey in these words:

"Sufficient data were secured to establish the presence of a gold belt 300 miles in length in Alaska, which enters Alaska near the mouth of Forty-Mile Creek, and extends westward across the Yukon Vallev at the lower ramparts. Its further extent is unknown. There is plenty of room for many more prospectors and miners in Alaska, for the gulches and creeks which have shown good prospects are spread over an area of seven hundred square miles. It is the opinion of the geologist in charge of the expedition that it is entirely practicable to prosecute quartz mining throughout the year in this region."

As far as any thorough prospecting is concerned, the basin of the Yukon has hardly been entered. The main river winds through a distance of more than 2,000 miles, and its tributaries vary in length from 60 to 300 miles, and, until the present season, not more than 5,000 miners had entered the basin. Hundreds and hundreds of square miles are absolutely unknown. The Yukon Basin contains an extent of territory fully equal to all the mining districts of the Pacific Coast and the Rocky Mountains put together. Accordingly, there are unbounded possibilities of further rich discoveries, because the conditions and the character of the entire basin are similar to the Klondike. No hardy, energetic man, who has the courage and determination to face the particular hardships and privations of prospecting and mining in Alaska's interior, need hesitate for fear that the field will all be taken.

There is very little opening in the Yukon mining fields for professional men, clerks, mechanics, etc., unless, of course, they have a good practical or theoretical knowledge of prospecting or mining.

Nor should a man start for the Yukon with less than $1,000 capital. Those who go should be prepared to stay a couple of years, as the average prospector has to sink a great many shafts through the frozen ground before he reaches anything worth his while. Ordinarily, during that time he has excellent chances of making money, with the possibilities of a fortune.

CLIMATE.

The climate of the Yukon Valley is one of extreme rigor in winter, with a brief but relatively hot summer, especially when the sky is free from clouds. The rainfall is small, averaging 10 inches, or about one-fourth that of Pennsylvania or other similar parts of this country. Winter sets in early in September and lasts until May. The lowest temperatures during the winter of 1889–90 were 32°, 47°, 59°, 55°, 45°, and 26° below zero in the months from November to April, and the highest summer temperatures are about 95°. The change from winter to summer is quick, on account of the rapid change in the length of the day, about seven minutes per day. The shortest days are about four, and the longest about twenty hours long. In the short winter days, the sun is so little above the horizon and so apt to be covered with clouds that, but for the bracing air, that season would be very depressing.

TIMBER.

A great part of the Yukon Valley is clothed with forests of spruce, cottonwood, and birch timber, all of which is good for building purposes. The hills are thickly covered with large trees, but the valleys have the best timber. A sawmill at Dawson City supplies that region with sawn timber at a cost of $130 per thousand feet, and next summer will see several others established where they will be needed.

GAME.

The country contains comparatively little game. Moose, caribou, and black, brown, grizzly, and white bears are occasionally seen. Swallows are quite thick in summer, and geese and wild ducks breed in large numbers on the rocks and rivers. The fishing is good, particularly for the king salmon, which weighs 80 to 100 pounds; grayling, white fish, lake trout, and eels also run up to large sizes. Swarms of mosquitoes and, later, of black gnats, are the pests of the Yukon lowlands in summer, and have even been known to drive the miner from his work.

AGRICULTURAL POSSIBILITIES.

There is a considerable area of land in the Yukon country with low valleys and good soil, which, with the influx of mining population, will be extensively cultivated. It is not a farming country, but can be successfully tilled for local supplies. Barley and oats mature, but potatoes have barely time to ripen. There are gardens at Forty-Mile and Fort Selkirk in which are raised potatoes, barley and oats, turnips, lettuce, radishes, and cabbages. The United States Government has appropriated $15,000 for an experimental farm next summer at the junction of the Yukon and Tanana Rivers.

COST OF LIVING.

The cost of living in the Yukon Valley is at present very high, as will be seen from the following recent quotations: Flour, $12 per hundred weight (following are the prices per pound); moose ham, $1; caribou meat, 65 cents; beans, 10; rice, 25; sugar, 25; bacon, 40; potatoes, 25; turnips, 15; coffee, 50; dried fruits, 35; tea, $1; tobacco, $1.50; butter, a roll, $1.50; eggs, a dozen, $1.50; salmon, each, $1 to $1.50;

canned fruits, 50 cents; canned meats, 75; shovels, $2.50; picks, $5; coal oil, per gallon, $1; overalls, $1.50; underwear, per suit, $5 to $7.50; shoes, $5; rubber boots, $10 to $15.

All miners unite in saying that the only fear for the coming winter is the lack of supplies. It is entirely probable that there will be so little supplies in the mining region this winter, in proportion to the number of men there, that prices will run much higher, and that great suffering and hardship will result temporarily, as it is impossible to get in fresh supplies after the winter has set in, *and provisions are sold only for cash.* But by next summer it is likely that ample means will be taken to ward off any such danger thereafter, and prospectors and miners can enter the gold region next spring without the fear of starvation or any greater hardships than naturally belong to a frontier region, even in such a comparatively inaccessible part of the globe as the Yukon Valley.

CHAPTER II.

HOW TO TAKE CARE OF YOURSELF.

MEDICINE CHEST.

We would advise all persons who contemplate going to the Klondike region to include in their outfits a medicine chest composed of the following drugs, which should be obtained at a drug store for about $5:

ACETANELID COMPOUND TABLETS (gr. v.).—To lessen fever, to relieve pain of neuralgic or rheumatic nature, for toothache or headache. Dose: Two tablets. May repeat in 4 or 5 hours. If 2 tablets do not relieve, 3 or 4 will not. Will usually relieve pain in chest—pleurisy pain.

MORPHINE (gr. $\frac{1}{8}$).—To relieve extreme pain of any sort; for instance, pain of broken limb or injury when it is unbearable, severe colic or pain in bowels—in fact, unbearable pain anywhere. Dose: One tablet. May repeat every 45 minutes until pain is relieved or person becomes drowsy. When it makes the person sleepy or drowsy, stop it.

DOVER'S POWDER TABLETS (gr. v.).—To produce sweat, break up a cold: Take two at bedtime with hot drink and wrap up warm. To check a bad diarrhœa: Take one every hour until diarrhœa is checked, or until drowsy.

ALOIN STRYCH. BELLAD. AND CASCARA PILLS.—Laxative. Two or three at bedtime for several nights.

COMPOUND CATHARTIC PILLS.—Physic, stronger than laxative. One or two at bedtime. A good plan is to take these one night, and then every night for a week take the laxative.

Bismuth Pepsin and Nux Vomica.—For indigestion or dyspepsia. One or two before each meal.

Bichloride Tablets, Compressed Antiseptic Tablets (Poison).—(These are poisonous if swallowed; the solution made from them is poisonous if swallowed). One dissolved in 1 pint of water to wash out any wound or sore that has pus or matter in it. Bathe for 5 minutes.

One dissolved in 1 quart of water for any fresh cut or sore. Bathe for 5 minutes.

After washing with this solution a sore, wound, or any place where the skin is broken, cover the place with five or six layers of iodoform gauze, right next the sore, and outside this plenty of cotton, and bandage or plaster to hold it in place.

Wounds or sores that are festering (forming pus or matter) should be dressed once or twice daily, according to the amount of discharge; those that are not forming matter or pus need not be dressed oftener than every three days.

Carbolic Salve.—A good dressing for burns, chafes, and small cuts and frost-bites.

The Best Liniment for Bruises.—Very hot water constantly applied, kept very hot, for 3 or 4 hours at a time

Citric Acid.—Dissolve in water to make lemonade in case of scurvy. A pinch to glass of water.

Lead Acetate, 12-Grain Opium Powders.—For lead and opium wash or liniment. It is made by boiling for 10 minutes 1 pint of water containing 1 heaping teaspoonful of lead acetate and 1 (12-grain) opium powder. Apply cotton saturated with this—hot, and change it as often as it cools; keep it hot (paper over dressing helps this). An excellent dressing or liniment for painful bruises or swellings, sprained joint, swollen testicle, painful bruise from fall, kick, or blow. It is not good for extensive raw surface as large cut or burn.

MONSELL'S SALTS FOR HEMORRHAGES.—In quantities in accordance with the person's liability to attacks of the trouble.

Two drams iodoform, 50 quinine pills, ½ dozen assorted bandages, 1 pound of listerine (for cuts, burns, colds, sore throat, etc.), 1 small roll of surgeon's plaster, and some antiseptic gauze dressing for wounds.

SURGEONS' LINT.—One yard.

ABSORBENT COTTON.—Four ounces.

MUSTARD PLASTERS.—One-half dozen.

MEDICAL SUGGESTIONS.

For many constitutions, the bracing effect of a trip to northern latitudes is positively beneficial. Snow and ice are not in themselves by any means injurious to the physical health of the average native of the temperate zone. They may be disagreeable, but they are not unhealthful, unless the soil of the district where they occur is of a nature to retain dampness. Clays are bad in this respect; gravelly soils are safe.

Scientific records have well established that the average duration of human life is greater in proportion as the residence is advanced from the equator towards the poles. There are exceptions, of course, but only such as prove the general rule. There is more risk of disease by far in a voyage to India or Panama than in one to Bering Straits.

Climate, however, is not the only thing to be considered, when there is question, in a medical sense, of the risks of a distant and laborious expedition undertaken by a multitude of persons, widely differing, as all multitudes must, in the capacity of individuals for standing hardship and privations.

To the weak, or those disposed to special ailments, conditions which are only invigorating to the man in average health are often absolutely fatal. Weak hearts and weak lungs can not face northern blasts or temperatures below zero. Rheumatism and its kindred affections are equally ill fitted for such tests. Nor are such persons, whether young or old, as have been long accustomed to purely sedentary occupations, or of lives of ease and luxury, physically fitted for the hardships of the Klondike. In the former, the vital and resistive powers have never been developed; in the latter they have been sapped. Weak eyes would be severely tested by the glare of a snow-covered land, and blindness is but one of the dangers to be feared by Arctic explorers.

In brief, we would say that persons subject to troubles of the heart, throat, or lungs should stay away from the Klondike. Physical exhaustion, colds, scurvy, rheumatism, and snow blindness are the chief dangers to be apprehended on the trip from a medical standpoint. For the healthy in other respects than those mentioned, there is no more danger to be dreaded than in any ordinary change of residence. Alaska is not in itself more unhealthy than Illinois, Norway, or the northern part of Scotland. Only those that are able to bear hardships should face them, but as far as hygienic conditions go, there need be no special apprehension on the score of Alaska. In conclusion, those of weak lungs and weak hearts, and sedentary people generally, are advised to stay at home. The others may balance their chances without need of doctor's advice.

CAMPING AND TRAVELING.

Establish camp rules, especially regarding the food. Allot rations, those while idle to be less than when at work, and also pro rata during heat and cold. Pitch the tent on top of the snow, pushing the poles and pegs down into it.

While some are busily engaged in building a fire and making a bed, let the best cook of the party prepare the supper. If you have no stove, build a camp-fire, either on an exposed point of rock or in a hole dug in the snow; if you have a stove, arrange it on a "gridiron" inside the tent, the gridiron consisting of three poles some six or eight feet long, and laid on the snow, on which the stove is placed. The heat from the snow will soon melt a hole underneath, but there will be enough firm snow under the ends of the poles to hold it up. For the bed, cut hemlock brush and lay it on the snow to the depth of a foot or more, and cover this with a large square of canvas, on which blankets and robes are put. When finished, it forms a natural spring bed, which will offer grateful rest after hauling a sled all day. In all except the most sheltered locations, the tent is necessary for comfort, and the stove gives better satisfaction than the camp-fire, and as it needs but little wood, is easier to cook over, and does not poison the eyes with smoke. There are fewer cases of snow blindness among those who use stoves than among those who crowd around a smoking camp-fire for cooking or warmth. Comfort in making a trip of this kind will depend, in a great measure, upon the conveniences of camping, suitable clothing, and light, warm bedding. Choose your bunk as far from the tent door as possible, and keep a fire hole open near your camp. If by any chance you are traveling across a plain (no trail) and a fog comes up, or a blinding snowstorm, either of which will prevent you taking your bearings, camp, and don't move for anything until all is clear again. Travel as much on clear ice as possible. Don't try to pull sledges over snow, especially if soft or crusty. White snow over a crevasse, if hard, is safe; yellow, or dirty color, never. Press the trigger of your rifle. Don't pull it. Don't catch hold of the barrel when 30 degrees below zero is registered. Watch out for getting snow in the barrel. If you do, don't shoot it out. Shoot a dog, if you have to, behind the base of the skull; a horse, between the ears, ranging downwards; a deer behind the left shoulder or in the head. If you can not finish your rations for one day, don't put back any part, but put it into your personal canvas outfit bag. You will need it later, no doubt. Don't waste a single ounce of anything, even if you don't like it. Put it away, and it will come handy when you will like it. If it is ever necessary to cache a load of provisions, put all articles next to the ground which will be most affected by heat, providing, at the same time, that dampness will not affect their food properties to any great extent. After piling your stuff, load it over carefully with heavy rocks. Take your compass-bearings, and also note in your memoranda some landmarks near by, and also the direction in which they lie from your cache; i. e., make your cache, if possible, come between exactly north and south of two given prominent marks, so that you can find it.

From the close of navigation by the freezing up of the lakes and rivers, the only means of travel is by dog trains. The dog used for this work is large, gaunt, long-haired, and wolfish, and will make 50 to 70 miles a day with a load that a man could scarcely haul at a walk. Six dogs make a good team, though a man and a dog or two dogs are often seen. The native packers often have as many as a dozen in a pack. The regular rule of the Hudson Bay packers is about 100 pounds to the dog, and six dogs in a team. The dogs weigh from 40 to 70 pounds. They are well trained and do not get sore feet, as do "tenderfoot" dogs taken into the country. They are fed on cheap bacon, horse meat, corn meal, or anything that is cheap and eatable. Dogs taken in from warmer climates are apt to get footsore, and tire out and be of no use. It requires an expert dog handler to get along with a dog team, and unless conditions are favorable, it is hardly worth while for a novice to make the experiment.

CARE OF FURS AND CLOTHING.

Keep your furs in good repair. One little slit may cause you untold agony during a march in a heavy storm. You can not tell when such will be the case. If your furs get wet, dry them in a medium temperature. Don't hold them near a fire. Keep your sleeping bag clean. If it becomes inhabited, freeze the inhabitants out. Keep all your draw-strings on clothing in good repair. Don't forget to use your goggles when the sun is bright on snow. A fellow is often tempted to leave them off. Don't you do it. A little dry grass or hay in the inside of your mitts, next your hands, will promote great heat, especially when it gets damp from the moisture of your hands. After the mitts are removed from the hands, remove the hay from the mitts and dry it. Failing that, throw it away. Be sure, during the winter, to watch your footgear carefully. Change wet stockings before they freeze, or you may lose a toe or foot.

When your nose is bitterly cold, stuff both nostrils with fur, cotton, wool, or anything to prevent the inside of your nose and throat from becoming frostbitten. The cold will cease. Don't eat snow or ice. Go thirsty until you can melt it. No man can continuously drag more than his own weight. In cases of extreme cold at toes and heel, wrap a piece of fur over each extremity. Remember success follows economy and persistency on an expedition like yours. And take warning—let no man go into the Yukon who can not endure hardship, who can not work with pick and shovel for ten hours, who can not carry a pack, and who can not take a full outfit with him.

CHAPTER III.

YUKON MINERS' OUTFIT.

PRELIMINARY SUGGESTIONS.

In addition to the medicine chest described in the preceding chapter, the following is a list of supplies necessary for one man for one year in the Klondike mining region. All authorities are agreed that those who go with less will be tempting fate. The requirements in clothing, food, and miners' outfitting in general are peculiar, and differ for the Yukon district from all other mining countries. The extreme long, cold winters, short, hot summers, and other points have to be considered. Outfits should be purchased in Tacoma, or Seattle, or other Puget Sound ports. The markets there offer everything mentioned below, in good quality and at reasonable prices. The merchants understand the trade and will select and put up an outfit, large or small, and unless a man knows what he wants, the best thing he can do is to name the price he can afford to pay and leave the selection to the merchant. This is possible, because there are manufactured the flour, bacon, evaporated fruits and vegetables, blankets, special Alaska clothing, boots, shoes, sleds, stoves, tents, boats, and, in fact, nine-tenths of the staples used in Alaska. Large stocks of goods are also carried at Juneau.

PROVISIONS.

One hundred and fifty pounds entire wheat flour; 50 pounds corn meal; 100 pounds hardtack; 50 pounds germea; 25 pounds rice; 75 pounds Bayo beans; 50 pounds peeled and evaporated potatoes; 10 pounds evaporated onions; 50 pounds dried soup vegetables; 5 pounds split peas; 50 pounds of stoned and dried fruit (apples, peaches, plums, prunes, apricots, and raisins); one hundred pounds boneless bacon; 50 pounds chipped beef; 5 pounds compressed soup and beef extract; 30 pounds canned lard; 30 pounds canned butter; 5 pounds coffee extract; 20 pounds compressed tea; 40 pounds loaf sugar; 15 pounds salt; 1 pound white pepper, ground; 7 pounds baking powder; 2 pounds soda; 6 packages yeast cakes; 1 pound mustard; ¼ pound ginger; 20 pounds condensed milk; 2 pounds evaporated vinegar or citric acid; 40 pounds candles; 7 pounds laundry soap; 5 cakes tar toilet soap; 1 pound of matches. In making purchases, it is well to observe the suggestion that the very best articles that can be purchased are none too good, and will more than repay the purchaser in the long run. Germea is selected in preference to oatmeal, because containing nutriment in a more condensed form, and because it is a very quickly prepared dish. Tea is far preferable in the winter cold than coffee.

SHELTER.

One 10 ft. × 12 ft. wall tent, made of 8-ounce duck or heavy drill, will make very satisfactory temporary quarters for four men. In lieu of a tent, a remarkably ingenious affair called a "Klondike Home" has been devised by Arthur F. Howes, of Seattle, Wash. These "homes," which cost $200, and weigh about 130 pounds, are built of aluminum. They are intended for two people, but, it is claimed, will house four and even six. It is said that the homes can be put up in an hour or less, and taken down and put in bundles in the same time. The frame is made of steel tubing, "telescoping" into the posts, which in turn are packed into the stove funnel, and this is put into a canvas cover for protection in transit. The material which forms the walls and roof is aluminum. The stove is of sheet steel, packed in a canvas bag to allow easy handling, and there are no sharp corners to cut into the carrier. It has two six-inch rounds on the top, and an oven of sufficient size, around which all the heat must pass. A copper-bottom boiler and a tea kettle are packed inside the fire-box, and there is room in this and in the oven to pack food and cooking utensils. The roof and walls of the "home" are packed into the bed, a sort of flat hammock, which when in use as a bed supports two people three feet above the floor. The frame is held together by connections resembling pipe fittings. This is covered with sheet aluminum, so lapped and fitted together that it is water and weather proof. It is guyed from the upper corners, like a tent. Hooks are provided on which to hang clothing, etc. A shelf along the side serves as a table. The bed is folded and used as a bench seat during the day. The windows are of transparent celluloid and are practically indestructible. The "homes" are 6 ft. 6 in. long × 5 ft. 6 in. wide × 6 ft. 6 in. high.

CLOTHING.

Those who have never wintered in the Arctic regions have no conception of the intense cold it is necessary to protect against. This lack of realization is doubtless responsible for the absolute insufficiency and uselessness of what is offered as a Klondike clothing outfit by some clothiers. It is winter that one must think of in preparing, not summer. For clothing, then, take the following: 3 suits heavy woolen underwear; 6 pairs heavy double-foot wool socks; 1 pair double-foot German woolen socks; 1 pair Canadian laragans or shoe-packs; 3 pairs seamless felt ankle moccasins; 2 pairs heavy snag-proof, leather soled, nail-protected, hip rubber boots, for summer work in water; 2 pairs specially made prospector's shoes; 2 heavy flannel shirts; 2 suits corduroy; 2 pairs pantaloons (one of stout, moderately heavy cloth for summer wear, the other of a lighter, close-woven cloth, which should be quilted with cotton batting a half-inch thick in the seat and over the thighs and knees, for winter use); 2 pairs stout riveted overalls, one felt-lined; 1 pair heavy Giant Buckle suspenders; 2 vests; 1 cloth sack coat; 2 heavy overall jumpers; 1 coat of wool cloth; 1 coat made of skin dressed with the hair on. The last named should be turned inside. For material, short hair, coarse wool, or sheepskin is excellent, and should not be expensive. For those who desire it, this coat can be made of deerskin with the hair outside and lining of squirrel or cat skins; 1 wool neck-scarf; 1 scarf or belt to draw coat tight around waist in cold weather; 2 pairs boots made of deerskin, hair outside, reenforced leather sole, sewed or pegged on, not nailed. The tops should extend above the knee and be laced tight there. They should be large enough to take in the foot covered with two pairs of socks, or with one pair and felt moccasins, and to take in the legs of the quilted pantaloons. The top of the boot need not be lined with fur. One heavy rubber-lined coat or mackintosh; 2 heavy woolen sweaters; 1 suit oil clothing; 2 fur caps, with fur-lined ear-laps; 1 wide-brimmed felt hat; 3 pairs heavy wool caribou mitts; 1 pair unlined leather work gloves; 1 pair snow-shoes; 1 pair snow-glasses; 6 towels; 2 dozen best quality bandana handkerchiefs; 10 yards mosquito netting, or 1 bee hat; 1 pair heaviest woolen blankets; 1 fur robe for sled travel and sleeping outdoors. This should be 4 ft. × 6 ft., with the outside covered with heavy woolen cloth and arranged like a bag by being made from a piece 6 ft. × 8 ft., doubled and sewed at each edge, except one. The end which is left open should be sewed about 6 inches toward the center from each side, and puckering strings arranged to draw the hole up close; 1 sheet light-weight rubber waterproof cloth 12 feet square; several rubber bags or packing cases should be purchased in which to carry perishable goods; compass, pocket comb, mirror, toothbrush, toilet paper, etc. In addition, there should be a small lot of repair materials—needles, thread, buttons, buckskin, shoemaker's awl and wax, boot-sole nails, rubber cement, rubber patching, etc.

HARDWARE, ETC.

One sheet-iron stove, folding, if possible, and pipe, the latter in flat sheets, with seam edge crimped for joining, or in three lengths, telescoping; 1 fry pan, with folding handle; 2 pots, 8-quart and 6-quart, with cover and bail; 1 kettle; 1 galvanized water bucket; 1 small riveted teapot; 3 pans for bread baking, sizes to nest together; 3 soup-plates, blue or granite ware; 2 cups, blue or granite ware, sizes to nest; 1 can opener; table-knife; fork; tea and soup spoons; 1 large mixing spoon; 1 bread or butcher knife. To save weight and for ease in cleaning, aluminum ware is very much to be preferred for the cooking utensils. A combined rifle and shotgun and ammunition; 1 large two-blade hunting knife; fishing tackle and hooks; 1 50-ft. tape; 1 gold pan; 1 gold scales; 2 3½-pound picks, with large eye; 3 handles for same; 1 drifting pick and handle; 1 long handle miner's shovel (spring point); 1 short handle miner's shovel (spring point); 1 scythe-stone; 1 pack strap; 1 American ax; 1 hatchet, hammer head, claw; 1 blacksmith hammer; 6 8-inch files and 2 taper files; 1 5½-ft. whipsaw, for getting out lumber; 1 26-in. Disston cross-cut handsaw; 1 rip handsaw; 3 chisels, including 1 calking chisel; 1 brace and bits; 1 folding draw-knife; 1 saw set and file; 1 square; 1 jack-plane; 20 pounds spikes; 2 pairs 8-inch strap-butts; 200 feet ½-inch manilla waterproof rope; 8 pounds of pitch; 5 pounds of oakum; nails, five pounds each of 6, 8, 10, and 12 penny;

plumb, level, chalk lines. In addition, each man in the party will require a Yukon sleigh, a skeleton affair made from the best hard wood and shod with ground brass runners. It is 7 feet 3 inches long and 16 inches wide, just the proper width to track behind snowshoes, and its cost is from $7 to $14. Brass is preferable to iron for the shoes, as it slides more easily through the fine, dry snow one finds in the early spring.

The list looks long and the bulk very considerable, yet there is not an unnecessary article in it. If, however, several men propose to travel together and work in partnership, only a small portion of the kitchen outfit and tools require duplication. Some game and fish may be taken, so that the use of the full quantity of subsistence suggested may not be required; but it is far safer to provide the full amount than to risk the success of getting game.

The total weight of the outfit is about 1,500 pounds, and the cost about $350. After purchasing his outfit and paying for his ticket to Dyea or Skagway, the would-be gold miner should have from $200 to $500 to pay incidental expenses during his first year in the Yukon Valley. One thousand dollars would be a far safer figure.

DUTIES ON OUTFITS.

Many different reports having gone out concerning the amount of exemption to miners which has been granted by the Canadian Government, it may be well to state that instructions have been issued by the Canadian Government exempting from all duty miners' blankets, personal clothing in use, and broken packages of provisions being used, also cooking utensils in use, and 100 pounds of food for the journey, charging ordinary customs duty on everything in excess of this amount. The duty on a $350 outfit will amount to from $50 to $70.

CHAPTER IV.

ROUTES TO THE YUKON GOLD FIELDS.

RECENT DEVELOPMENTS.

Gold, and the search for gold, ofttimes make a wonderful change in the face of nature. Before the rich discoveries on the Klondike, Alaska was a land of slow-going old settlements, typical of all that was leisurely and ancient. A few steamers starting from Tacoma, the head of navigation on Puget Sound, and stopping at Seattle, Victoria, Vancouver, and other Puget Sound ports, sufficed to take care of the trade in supplies, furs, fish, and occasional prospecting parties, and made up the life of the community at the extreme southerly edge and only inhabited portion of that vast territory.

With the spreading of the news of the discovery of gold on the Klondike River began to come the crowds of eager gold seekers, by hundreds and then by thousands, from every quarter of the world, and at the principal Puget Sound ports, instead of the comparatively uneventful sailing of a steamer about once a week to care for the traffic in supplies, ores, fish, furs, and the few travelers back and forth, with the added interest in the summer months of the tourist excursions, now the docks present a scene of bustle and excitement. The crowds of gold seekers, with their outfits, their pack ponies, or dogs, the friends who go down to bid them good-by, and the rush and hurry in getting aboard the vessel's freight, make quite a different picture from the sailing of an Alaskan steamer a year ago. Along the streets of the cities are displayed "Klondike" outfits, Alaska clothing, sleds, sleeping bags, miners' tools, condensed foods for the

Arctic regions, gold pans, rifles and revolvers, dust belts, and the many other articles that go to make up a prospector's outfit, all looked upon with interest by the passing crowds in the streets. In the shops, workmen are busy building Yukon sleds, camp stoves, and other equipment for the prospector, and the stamp of "Klondike" is seen upon every line of industry. Parties are flocking into the cities by every train, to await the time to start for the north in February or March, and groups of them may be seen talking over the various fields, the prospects and chances, and the experienced miner giving the tenderfoot points on outdoor life. At Juneau, many of those who failed to get over the pass in the fall are waiting for the spring, and at Dyea, Skagway, Fort Wrangel, and Sitka, others are awaiting the lengthening days of February to be off to the Klondike, the Copper River, the Pelly, the Tanana, the MacMillan, the Munook, and the dozen other fields where rich finds are reported, and the first of March will see a rush of gold hunters into the country such as has seldom been seen in any former gold-mining excitement in the history of the world.

THE FIRST OBJECTIVE POINT.

As nearly all Alaskan steamers sail from Puget Sound ports (Tacoma, Seattle, Victoria, Vancouver, etc.), one of these cities will be the first objective point. If the would-be prospector is a citizen of the United States, he will probably go to either Tacoma or Seattle, which are near neighbors, both in the State of Washington; if he be a Canadian citizen, he will naturally go to Victoria or Vancouver.

Rates from New York City to Puget Sound points vary from $62.75, the lowest second-class rate, to $81.25 for first-class. Meals and berths are not included in these figures. Berths from New York to Puget Sound points are $9. The

Union Pacific, Northern Pacific, Great Northern, and Canadian Pacific lines are the principal competing roads. All of these lines have offices in New York. The *cheapest* way of getting from New York to the Pacific Coast is by steamer from New York to New Orleans, thence by the Southern Pacific to San Francisco. The cost of a ticket by this route, including steerage berth and meals on steamer and second-class passage by rail, is $54.50.

From Puget Sound points there are at present nine known routes to the Yukon gold fields.

As the Chilkoot Pass route is the most important and the most generally traveled of these routes, we will describe that first, reserving the description of the other routes for a subsequent chapter.

THE CHILKOOT PASS ROUTE.

The shortest route to the basin of the Yukon, and the one which has been taken by nearly nine-tenths of all the gold seekers who have thus far gone to the interior, is that via the Chilkoot Pass (see map). The details of this route are as follows:

From Puget Sound ports to Juneau, 900 miles.—This portion of the journey can be taken at any time of the year. There are usually several steamers each way every week between Puget Sound ports and Juneau. The trip usually takes five or six days. Rates at hotels and restaurants in Juneau are about the same as in any city. The fare from Puget Sound ports to Juneau, including berth and meals, is, first-class, $32; steerage, $17. One hundred and fifty pounds of baggage are allowed each passenger, and excess baggage is carried at $9 per ton. In the summer season, passage can be obtained direct from Puget Sound ports to Dyea, without changing at Juneau.

From Juneau to Dyea (or Taiya), 101 miles.—Ordinarily,

passage over this portion of the route can be obtained within a day or so after arrival in Juneau; but as the small boats run irregularly, quick connections can not always be depended upon. The fare from Juneau to Dyea is $8 to $10, and if the weather is fair and the load light, the trip is made in twelve hours. The landing process at Dyea is long and tedious. There is no deep water near shore. The tide at the head of the long and narrow estuary rises and falls twenty-three feet, and the beach is long and flat. Hence, everything must be taken ashore in lighters and surf-boats, which make long trips with each load. Horses are dumped into the water to wade ashore. The responsibility of the steamship company ends at the anchorage, but it uses its boats and crews to help get passengers and freight ashore. As soon as possible after landing, the freight is sorted and carried out of reach of tidewater. Most miners camp near by in the edge of the woods, perhaps taking one or two meals at a restaurant; others find both board and lodging until they are ready to push on. Now, for the first time, the miner begins to size up his belongings, and begins to realize that a proper outfit for a trip of this kind is the result of experience, and the longer he has been in this country and the more thoroughly he knows it, just so much more care is used in the selection and packing of his outfit. A careful and thorough examination should be made to see that nothing has been lost or forgotten. Towns of five thousand inhabitants have grown up both at Dyea and at Skagway, five miles below.

OVER THE CHILKOOT PASS—THE NEW WAY.

The problem of how to rapidly and cheaply transport passengers and freight over the Chilkoot Pass to the headwaters of the Yukon has been solved in an unexpected but entirely practical manner. Heretofore, getting over this pass has been the most dangerous and difficult part of the Yukon journey. Men have had to either carry their heavy outfits a distance of twenty-five miles over the pass, involving lugging them up steep hills with an aggregate elevation of 3,500 feet, or pay large sums to the Indians for taking them over. This has required a great deal of hard work when the miner's time was most valuable, or necessitated the expenditure of such large sums in getting over that few men could afford it. Men who have done their own packing have been tired and worn out on reaching the lakes, and some have given up on reaching Lake Lindeman.

This situation will be quite changed by the Chilkoot Railroad and Transport Company, which expects to have its transportation system in operation by February, 1898, in time for the early spring rush over the pass. Miners and prospectors can then disembark from the steamers at Dyea, and in twenty-four hours find themselves with their outfits at Lake Lindeman. Boats, sleds, and dogs can be taken over as readily as flour and beans, so that on reaching Lake Lindeman the miner has only to pack his sled, launch his boat, or harness his dog team, and be ready to start down the lakes. The great majority of miners will have neither dogs nor boats, but on reaching Lake Lindeman will proceed to haul their sleds over the ice to the foot of Lake Lebarge, where they will build boats and be ready to start down the river as soon as the ice breaks.

The Chilkoot Railroad and Transport Company is now (December, 1897) completing a railroad eight miles in length from Dyea to the mouth of Dyea Canyon. Between the mouth of the canyon and Crater Lake, on the other side of the summit, two aerial tramways, each four miles in length, will be operated. The contract for these aerial tramways has been let to the Trenton (New Jersey) Iron Works, which has built over a hundred of them now in successful operation in the United States, Mexico, Central and South America. The longest one in the world, twelve miles in length, was recently built by the same company in the island of Hayti.

One tramway on Chilkoot Pass will reach through Dyea Canyon to Sheep Camp, with a rise of 1,000 feet in four miles. The other will extend from Sheep Camp to Crater Lake, with a rise of 2,500 feet to the summit of the pass, and a decline of 500 feet between the summit and Crater Lake. Over 16 miles of the best steel-wire cable will be used in the construction of these tramways, which will be supported by solid iron supports placed every 100 feet. The power station, to develop 50 horsepower, will be located at Sheep Camp. This power will operate the tramways rapidly, giving them a daily capacity of 120 tons, or outfits for 200 men. Special carriages will be provided for carrying passengers. The Trenton Iron Works is under contract to have this system of tramways in operation by January 15, and beginning then the Chilkoot Railroad and Transport Company will be able to transport 200 miners and their outfits over the pass daily. From Crater Lake to Lake Lindeman Landing the trail runs down hill, with an easy grade, making it possible for the miners to load their sleds and slide down over the crusted snow and ice.

A telephone line connecting Dyea and Lake Lindeman is being constructed, ensuring the operation of the transportation system to the best advantage. It will also enable men crossing the pass to communicate with either end or intermediate stations.

OVER THE CHILKOOT PASS—THE OLD WAY.

From Dyea to the Sheep Camp, 12 miles.—Two days are generally consumed in making this part of the journey, although it is possible to make it in one. The first five miles of the journey, from Dyea to the mouth of the canyon, can be made by canoe during parts of April and September and all of May, June, July, and August, or on the ice at other seasons of the year. Dyea Canyon is about two miles long, and can be traversed on the ice in winter by bridging the dangerous holes with poles. In summer it is necessary to go around the canyon by a trail which has been built on the east side. The balance of the journey is easy. There is now a considerable settlement at the Sheep Camp.

From the Sheep Camp to Lake Lindeman, 15 miles.—This portion of the route, containing the Chilkoot Pass, can be traversed in summer by any able-bodied man or woman with very little difficulty or danger, but the conditions in winter, spring, or fall are such that great caution must be observed. This is due to the terrible severity of the Arctic storms, which come up suddenly and rage with fury through this narrow defile at certain seasons of the year. During the month of October, 1897, not less than 20 persons are reported to have perished in such storms. None except those who are familiar with the pass should ever attempt it alone. Four or five men should compose each party starting for the gold fields, as one tent, stove, set of tools, etc., will suffice, and the hardships can in many respects be lightened by cooperation on the part of all. The Sheep Camp is near the summit, and no wood for a fire can be gotten until timber is reached on the other side of the pass, about three or four miles from the head of Lake Lindeman. For this reason, the Sheep Camp is not usually left until all of the outfit has been placed on the summit. When the weather is favorable, everything except what is necessary for camp is pushed a mile and a half to Stone House, a clump of big rocks, and then to what is called the Second Bench. From this point, for about 600 feet, the trail is very difficult, but Indians may be hired to carry packages up it for about $5 per hundred pounds. At most seasons of the year, horses can be used for the purpose, if sharply shod and accustomed to mountain trails. A good horse will carry 150 to 200 pounds over the summit, while for the average man 50 pounds is quite sufficient. The descent for the first half mile is steep, then a gradual slope to Lake Lindeman, some ten miles away. But there is but little time for resting and none for dreaming, as the edge of the timber where the camp must be made is seven miles from the summit. Taking the camping outfit and sufficient provisions for four or five days, the sleigh is loaded, the rest of the outfit is packed up, or buried in the snow, shovels being stuck up to mark the spot. This precaution is necessary, for storms come suddenly and rage with fury along these mountain crests. The first half mile or more is made in quick time; then, over six or seven feet of snow, the prospector drags his sleigh to where there is wood for his camp-fire. At times this is no easy task, especially if the weather be stormy, for the winds blow the new-fallen snow about so as to completely cover the track made by the man but little ahead; at other times, during fine weather, and with a hard crust on the snow, it is only a pleasant run from the pass down to the first camp in the Yukon Basin.

DOWN THE LAKES TO THE HOOTALINQUA.

From Lake Lindeman to Lake Bennet, 5 miles.—The trip through Lake Lindeman is short, the lake being only 4½ miles long. Boats may be hired to carry goods across. It is necessary to portage from Lake Lindeman to Lake

Bennet, the portage, however, being less than a mile. The best time to reach Lake Bennet is early in the spring, say about April 15th, before the ice begins to melt. Persons who time their trip so as to reach there about that date can make the trip across the lakes on the ice, and need not build rafts until they reach open water on the Yukon or Lewis River. Those who reach Lake Bennet after the ice has melted will need to build or purchase a boat or scow at this point before proceeding further. It is necessary that one of the party should have a knowledge of boat building, for it is absolutely essential that the craft shall be stanch and substantial. The double-ended batteau is the pattern ordinarily preferred, though the plain scow of good depth is more easily built and can be depended upon. The boats are usually 22 to 24 feet long and 4½ to 5 feet wide, and, if purchased at the local sawmill, cost about $60.

From Lake Bennet to Lake Takou (or Tagish), 29 miles.—Lake Bennet is 26 miles long. Upon this lake the British boundary is crossed. Between Lake Bennet and Lake Takou is the Caribou Crossing, which is about 3 miles long and quite difficult.

From Lake Takou to Miles Canyon, 66 miles.—Seventeen miles of this distance is spent in crossing Lake Takou (or Tagish, or Takish, as it is variously spelled), 5 miles in descending the Six-Mile River, 19 miles in crossing Mud Lake and Marsh Lake, and the balance in descending the Fifty-Mile River. About half way down the Fifty-Mile River is Miles Canyon. Before reaching the canyon, a high cut bank of sand on the right-hand side will give warning that it is close at hand. Good rivermen have run the canyon safely, even with loaded rafts, but it is much safer to make a landing on the right side and portage the outfit around the canyon, three-quarters of a mile, and run the raft through empty. The sameness of the scenery on approaching the canyon is so marked that many parties have gotten into the canyon before they were aware of it.

From Miles Canyon to Lake Lebarge, 31 miles.—From Miles Canyon to White Horse Rapids, 2 miles, the boat or raft can be towed with safety down the stream, when a large sign will be seen on the left bank, with the words "Look

MILES CANYON.

Out" in letters a foot square. The White Horse is the most dangerous portion of the trip. It is a box canyon about a hundred yards long and fifty in width, a chute through which the water of the river, which is 600 feet wide just above, rushes with maddening force. But few have ever attempted to run these rapids, and fourteen of them are known to have been drowned. It is much safer to portage the outfit around these rapids and send the boat through empty. Even then the boat is likely to be lost or damaged. Every man's life should be worth more to him than all the gold in the Klondike region. The balance of the trip from White Horse Rapids to Lake Lebarge is via the Fifty-Mile River. It may be necessary, below the White Horse Rapids,

WHITE HORSE RAPIDS.

to build a new boat before the journey can be continued. It is probable that ere long a railroad will be constructed around Miles Canyon and White Horse Rapids, to facilitate the transportation of freight and passengers at this point. At present there is a portage road on the west side, and rollways in some places on which to shove the boats over.

From Lake Lebarge to Hootalinqua River, 61 miles.—There are no rapids or other dangers in this part of the journey. In the middle of Lake Lebarge (which is 31 miles in length) there is an island where parties bound for the gold fields usually camp. Ice on Lake Lebarge is usually good until about the last of April. From Lake Lebarge to the Hootalinqua, the course is down Thirty-Mile River.

DOWN THE YUKON VALLEY.

From Hootalinqua River to McCormick Trading Post, 111 miles.—This section of the journey is always made in summer, as the river never freezes over smooth. The ice freezes in great rough masses, which makes traveling well-nigh impossible. If a person should have the misfortune to get frozen in, the best thing he can do is to go into winter quarters and commence prospecting at once on the little streams in his immediate neighborhood; the entire region is known to be gold-bearing, and a lucky strike is almost as likely to be made in one place as another. From Hootalinqua River to the Big Salmon River is 27 miles; from the Big Salmon River to the Little Salmon River is 63 miles, and from there to McCormick Trading Post is 21 miles. The Hootalinqua River, after its junction with the Big Salmon River, is known as Lewis River.

From McCormick Trading Post to Fort Selkirk, 79 miles.—Twenty miles below McCormick Trading Post are the Five Finger Rapids. Here four large buttes divide the water in five passages; the right-hand passage is the only one which is practicable, and, though the water is swift, it is safe if the boat be kept in the center. A few moments of strong pulling and careful management, and the boat is rapidly approaching Rink Rapids, three miles below. Here again the right-hand side ensures safety, and having gone through them the last dangerous water is passed. Next comes the Pelly River, upon which, and especially upon its chief branch, the MacMillan, valuable discoveries were made late in 1897. The junction of the Pelly and Lewis forms the Yukon proper. Fort Selkirk, or Harper's (as it is sometimes called), is located at this point.

From Fort Selkirk to Stewart River, 106 miles.—Ninety-six miles below Fort Selkirk the White River is passed, 10 miles beyond which is the Stewart River, where some of the latest discoveries have been reported. There is little doubt but that the newcomer stands as good or better opportunity in prospecting along the creeks entering this river as anywhere in the Yukon district. All of the streams thus far referred to enter the Yukon from the right, with the exception of the White River. Below the Stewart River, the various mining districts are reached in the following order: Sixty-Mile Creek, Klondike River, Forty-Mile Creek, and Birch Creek. Gold has also been discovered on Porcupine River, Munook Creek, Tanana River, and Koyukuk River. An examination of the map will show that these districts cover practically the whole known interior of Alaska, and large deposits have been found on the Kenai peninsula and other places along the southern coast. Just at present the Copper River district (see map) is attracting much attention.

CHAPTER V

ROUTES TO THE YUKON GOLD FIELDS.

(Continued.)

THE ALL-WATER ROUTE.

One of the best, because safest, most natural, most comfortable, and cheapest routes to the gold fields is the all-water route (see map). From Puget Sound the steamers sail out through the Straits of San Juan del Fuca northwesterly across the Pacific Ocean 2,000 miles to Dutch Harbor, on Unalaska Island, which is the first stop; thence 750 miles north through Bering Sea and Norton Sound to St. Michael Island, 60 miles above the mouth of the Yukon, where transfer is made to the smaller craft which ply up and down the Yukon. The Yukon is navigable the entire distance from its mouth to Fort Selkirk, in Canadian territory, 2,300 miles, without a break, and all of the gold fields thus far discovered are reached direct by the river steamers. The fare from Puget Sound to the gold fields, meals and berth included, is only $200 first-class. Competition may reduce this to $150, or even $100, for ordinary second-class accommodations. Steerage passengers must furnish their own bedding. As all the streams in the immediate neighborhood of the Klondike have already been taken up (see map for particulars), persons of limited means are advised to wait until spring and go by this route. In doing so, they will pass both Circle City and Forty-Mile, near which some of the best placer mines on the American continent have been discovered, and were being worked at great profit before they were abandoned in the great rush for the Klondike. These districts are located entirely within American territory. Millions of dollars' worth of gold will undoubtedly be taken from these at present abandoned mining districts within the next few years. Navigation on the Yukon closes in September and opens in June. As the head waters are much farther south than the mouth, the break-up begins in the upper river and tributaries, and the ice, which freezes 5 feet thick in winter, packs and crushes its way towards the deltas. Navigation between Circle City and Dawson is possible sometimes a month before steamers can enter the mouth of the Yukon from St. Michael's on the upward trip.

One great disadvantage of going by the all-water route is that the best part of the season is gone before one can reach the mining fields.

The boats which ply up and down the Yukon calculate on making only two round trips during the season, and sometimes make but one. Two hundred passengers who left Puget Sound as early as July 22, 1897, for Dawson, via St. Michael's, were stranded on the Yukon flats, five hundred miles below Dawson, and had to turn back. The lightest draft boats carrying provisions could not get up the river in August and September because of the low water. Several thousand men who started in July and August via St. Michael's were stranded for the winter on that cold and desolate island or a short distance up the Yukon. Several hundred, however, were fortunate enough to reach Rampart City, at the mouth of Munook Creek, half way up the Yukon, and as rich strikes have been made in that neighborhood they are perhaps as well off as though they had reached Dawson. The gold here is heavy and coarse, running $18 and $19 to the ounce. The mines begin twenty-four miles up Munook Creek from Rampart City. In September, Munook claims were selling for $6,500 each. Gold was plentiful, and it was believed that winter work would prove them very rich. Gold has also been found on the Tanana River, a short distance below the Munook. The Tanana is a large river with many tributaries, offering a large district in which to prospect. It is 900 miles from St. Michael's to the mouth of the Tanana, 80 miles from there to the Munook, 450 miles from Munook to Circle City, 240 miles from Circle City to Forty-Mile, and 52 miles from Forty-Mile to Dawson.

The first vessel sailing in the spring from Puget Sound for St. Michael's leaves in April or May. The steamers start in May and June, as the Yukon River does not open at its mouth until the latter month.

Several parties have formed to take flat-bottom river boats to the Yukon in the knock down, and set them up there. A flat-bottom river steamer can be built for $1,000 to carry a party of ten with their provisions and supplies. Boats to carry parties of 100 to 200 cost $8,000 to $14,000.

THE WHITE PASS ROUTE.

The White Pass route starts from Skagway City, which is five miles below Dyea. From the harbor, where there is a long wharf to deep water, the trail follows the Skagway River to its head, which is near the summit of the pass, a distance of 16 miles. The first four miles are in the bed of the river, and the ascent is gradual. At four miles the canyon is reached, and here the route becomes more difficult. For seven miles the trail works its way along the mountain side, rising steadily for almost the entire distance. This is the only hard part of the route. The next three miles is a gentle rise, and they carry the trail to the summit, an elevation of 2,600 feet above the sea-level. This is 1,000 feet lower than the summit of the Chilkoot Pass, but, because of the impassable condition of the trail, has been quite disappointing to those who have tried to go over it. Surveys are being made for a railroad across White Pass, which will undoubtedly be built next year.

THE COPPER RIVER ROUTE.

The Copper River route is by steamer from Juneau to Valdes Inlet on Prince William Sound. From Valdes the passage to the Copper River is in a northeasterly direction over a dead glacier, which may be readily traversed during most of the year. The Indians living on the upper waters of the Copper River come out by that route on their semi-annual trading trips. The trail strikes Copper River about 100 miles north of its mouth, thereby avoiding the rapids, canyons, and glaciers that have prevented the successful passage of the river. It then proceeds up the Chittyna, the main branch of the Copper River, crosses the divide at Scoloi Pass, and goes down the White River to the Yukon. Some claim that the building of a railroad to the Yukon is entirely feasible by the Valdes route from Prince William Sound, and surveys are to be made in 1898.

One advantage of going by the Copper River route is that very good diggings are reported to have been found along the stream, some prospectors going so far as to state that the placers of Copper River are as rich if not richer than Klondike. Some of the Copper River Indians, however, are savage and well armed. They resent the coming of whites to take their gold. Frequently white miners are driven out of the country by the Indians, some of whom, particularly the Stik tribe, are mixed with Russian Finns and are vicious and aggressive. They are good shots, and, besides, from their places of vantage in the canyons they can resist a large body of white men. Several parties are organizing to explore the Copper River in the spring of 1898. The members of these parties are all well armed. One of their number saw, while passing through the Copper River country, squaws with bracelets of virgin gold which had been beaten out of nuggets.

THE DALTON ROUTE.

The Dalton trail leaves tidewater at Chilkat, a landing to the west and north of Dyea, and runs over a comparatively easy pass to the north of Chilkoot Pass, continuing to the westward of the lake country and striking the Yukon above Fort Selkirk, a distance from Chilkat of about 400 miles. This is the usual route for driving in beef cattle, as the pasturage, May to September, is good all the way over to the river. In taking in beef cattle over the Dalton trail, some prospectors have placed light packs on the cattle and succeeded in that way in getting their supplies in. Beef cattle costing $20 at Tacoma have been sold at Dawson City at from $700 to $900, where beef was worth 75c. to $1 per pound.

It is believed by many that this will ultimately be a popular route to the mines. It will be made passable for wagons as early as practicable in the summer of 1898, and it has even been proposed to construct a railroad along this entire route. Gold has been discovered on this route, about midway between Chilkat Pass and Fort Selkirk.

THE CHILKAT ROUTE.

This route is up the Chilkat Inlet, which enters the Lynn Canal a few miles below Dyea, thence up the Chilkat River and over Chilkat Pass (which is said to be about 1,000 feet lower than the Chilkoot Pass) and down the Tahkeena River to its junction with the Fifty-Mile River below White Horse Rapids. As the Chilkat Pass is 25 miles longer than the Chilkoot Pass, it is not much used at the present time.

THE STICKEEN ROUTE.

Goods and passengers intended for this route have to be transshipped from ocean-going steamers to river steamers at Fort Wrangel. The Stickeen, under favorable circumstances, is navigable for stern-wheel steamers of light draft and good power to Telegraph Creek, 140 miles from its mouth. The river usually opens for navigation between April 20th and May 1st. On the lowlands there is good grazing for horses and cattle from April 20th to about December 1st. The distance from the Stickeen at Telegraph Creek to Teslin Lake, the source of the Hootalinqua River, is about 120 miles. The portage is through a partly open and partly wooded country, somewhat rolling but not rough. A pack trail runs from Telegraph Creek to the head of the lake. At the head of Teslin Lake there is plenty of timber for whipsawing lumber to build boats for the voyage down the river to Dawson, or lumber may be purchased at the small sawmill now in operation there. From the Hootalinqua, the balance of the journey to the gold fields is by the same route as that described under the head of Chilkoot Pass route. From the head of Teslin Lake to the Klondike is 584 miles.

THE OVERLAND ROUTE.

This route starts from Ashcroft, B. C. The distance to Stuart Lake can be traveled with ease by pack train. No feed need be carried for horses, as there is an abundance of grass the entire distance. From Fort James (on Stuart Lake), the route is to Telegraph Creek, over prairie country. From this point the balance of the journey is the same as via the Stickeen route. If desired, the journey from Telegraph Creek to the Yukon may be made by the somewhat lengthy route via Dease Lake, Frances River, and Pelly River, which is said to be very easy.

THE TAKOU ROUTE.

This route is from the head of Takou Inlet, a little south of Juneau, overland by the valley of the Takou River to Lake Teslin, from which the balance of the journey is the same as via the Stickeen route. The gold fields can also be reached by the way of Edmonton, on the Canadian Pacific Railway, thence via the McKenzie, Frances, and Pelly Rivers.

CHAPTER VI:

MINING REGULATIONS OF ALASKA.

The misunderstanding and contentions regarding the laws that are applicable to Alaska, so far as the lands and claims are concerned, have been set at rest by a statement made by Commissioner Hermann, of the General Land Office. Many inquiries on this question have come to the Interior Department, and numerous applications have been made for copies of the Public Land Laws, which, however, do not apply to Alaska. All this is due to the gold boom. The General Land Office officials have taken much interest in the reports that come from the gold belt, and have investigated the laws that govern them.

Commissioner Hermann says these laws are applicable:

VARIOUS LAWS.

1. The Mineral Land Laws of the United States.
2. Town site laws which provide for the incorporation of town sites and acquirement of title thereto from the United States Government to the town site trustees.
3. The law providing for trade and manufactures, giving each qualified person 160 acres of land in a square and compact form.

The coal land regulations are distinct from the mineral regulations or laws, and the jurisdiction of neither coal laws nor public land laws extends to Alaska, the Territory being expressly excluded by the laws themselves from their operation. The Act approved May 17, 1884, providing for civil government of Alaska, has this language as to mines and mining privileges:

MINING CLAIMS.

"The laws of the United States relating to mining claims and rights incidental thereto shall, on and after the passage of this act, be in full force and effect in said district of Alaska, subject to such regulations as may be made by the Secretary of the Interior and approved by the President, and parties who have located mines or mining privileges thereon under the United States laws applicable to the public domain, or have occupied or improved or exercised acts of ownership over such claims, shall not be disturbed therein, but shall be allowed to perfect title by payments provided for."

There is still more general authority. The Act of July 4, 1866, says:

"All valuable mineral deposits in lands belonging to the United States, both surveyed and unsurveyed, are hereby declared to be free and open to exploration and purchase, and lands in which these are found to occupation and purchase by citizens of the United States, and by those who have declared an intention to become such under the rules prescribed by law, and according to local customs or rules of miners in the several mining districts, so far as the same are applicable and not inconsistent with the laws of the United States."

The patenting of mineral lands in Alaska is not a new thing, for that work has been going on as the cases have come in from time to time since 1884.

ALASKA MINING CODE.

The United States mining code applies in the Territory of Alaska, but miners' law, which agrees with the Federal law, has prevailed, pending the appointment of the proper officials to carry out the full provisions of the statutes. The general law, covering quartz and placer claims, condensed into brief space, is as follows:

QUARTZ DEPOSITS.

Quartz Land.—Mining claims upon ledges or lodes of precious metals can be taken up along the vein to the length of 1,500 feet and 300 feet each side of the middle of the vein. To secure patent, $500 worth of work must be performed and $5 an acre paid for the land—twenty acres. Six months' failure to do work forfeits the claim.

PLACER CLAIMS.

Placer Land.—Claims usually called "placers," including all forms of deposit, excepting veins of quartz or other rock in place, are subject to entry and patent. No single individual can locate more than twenty acres of placer land, and no location can be made by any company composed of no less than eight bona-fide locators, exceeding 160 acres. The price per acre of placer claims is $2.50. Where placers contain veins or lodes, the cost per acre is $5.

It is important that locators accurately mark and describe their claims. In marking, the locator may do so in any direction that will not interfere with the rights or claims existing prior to his discovery. Litigation, expense, and delay may be avoided by being particular in the matter of boundaries. The essentials are:

First—That the corners should be marked on the ground by stakes in mounds of earth or rock, or by marked trees or other natural objects.

Second—The notice of location should describe these corners so that they can be identified on the ground by the description, and, in addition, the direction and distance of one of the corners from a Government corner (if surveyed) or well-known object, such as a junction of roads, trails, or ravines, a bridge, building of any kind, or natural feature, as rock, etc.

PENALTIES.

It is a felony to sell a salted mine; to fraudulently change samples or assays with intent to defraud; to make or give false assay or sample with intent to defraud; to rob vein, sluice-box, quartz mill, etc., or trespassing upon mining claim with intent to commit a felony.

It is a misdemeanor to deface, tear down, or destroy a post, monument, boundary mark, or location notice; or without authority to take water from any ditch, pipe, reservoir, etc., or to unlawfully interfere with the same.

CHAPTER VII.

MINING REGULATIONS OF N. W. CANADA.

The following is a summary of the principal regulations made by the Canadian Government regarding placer mining along the Yukon River and its tributaries in the Northwest Territories, corrected to August 21, 1897:

In these regulations, "bar diggings" mean parts of a river which are flooded at high water but are not covered at low water. Mines on benches are called "bench diggings." "Dry diggings" mean mines over which a river never extends. "Claim" means the right to a diggings during the time for which the grant is made. A claim does not include any special surface rights. "Legal post" means a stake or stump standing four feet above the ground, and squared on four sides for at least one foot from the top. Both sides so squared must measure at least four inches across the face. "Close season" means the period during which placer mining is generally suspended, and is fixed by the Gold Commissioner in whose district the claim is situated. "Locality" means the territory along a river (tributary of the Yukon River) and its affluents. "Mineral" includes all minerals other than coal. Claims are measured horizontally, irrespective of the surface of the ground. Miners must be over 18 years of age.

BAR, DRY, AND BENCH DIGGINGS.

Bar diggings are strips of land 100 feet wide at high-water mark, and thence extending into the river to its lowest water-level. Their sides are formed by two parallel lines run as nearly as possible at right angles to the stream, and must be marked by four legal posts, one at each end of the claim at or about high-water mark, also one at each end of the claim at or about the edge of the water. One of the posts at high-water mark must be legibly marked with the name of the miner and the date upon which the claim was staked. Dry diggings and bench diggings are 100 feet square and are staked in the same way.

CREEK AND RIVER CLAIMS.

Creek and river claims are 100 feet long (discoverers of new mines or of new auriferous strata, in a locality where the claims are abandoned, are entitled to claims of double this length) measured in the direction of the general course of the stream, extending in width from base to base of the hill or bench on each side, but when the hills or benches are less than 100 feet apart, the claim may be 100 feet in depth. The sides consist of two parallel lines run as nearly as possible at right angles to the stream, and must be marked with legal posts at the edge of the water and at the rear boundaries of the claim. One of the posts at the stream must be marked as above.

Entry is granted only for alternate claims, the other alternate claims being reserved for the Government, to be disposed of at public auction. The penalty for trespassing upon a claim reserved for the Government is the forfeiture of all mining rights.

RECORDING CLAIMS.

A claim must be staked out as soon as located, and must then be recorded with the Gold Commissioner in whose district it is situated, within three days, if it is within ten miles of the Commissioner's office. One extra day is allowed for making such record for every additional ten miles. An entry fee of $15 is charged the first year, and an annual fee of $15 for each of the following years. In addition, a royalty of ten per cent. on the gold mined is collected, and in case the amount taken from any single claim exceeds five hundred dollars per week, there is collected upon the excess a royalty of twenty per cent. Default or fraud in payment of the royalty is punished by cancelation of the claim.

After the recording of a claim, the removal of any post for the purpose of changing the boundaries of the claim acts as a forfeiture of the claim.

The entry of every holder of a grant must be renewed and his receipt relinquished and replaced every year, the entry fee being paid each time.

No miner can receive a grant of more than one mining claim in the same locality, but the same miner may hold any number of claims by purchase, and any number of miners may unite to work their claims in common upon such terms as they may arrange, provided such agreement be registered with the Gold Commissioner and a fee of $5 paid for each registration.

Any miner may sell, mortgage, or dispose of his claim, provided such disposal be registered with, and a fee of $2 paid to, the Gold Commissioner. The Gold Commissioner may grant to holders such right of entry upon adjacent claims as may be necessary for the working of their claims, upon such terms as may to him seem reasonable. Every miner is entitled to the use of a share of the water naturally flowing through or past his claim.

A claim is deemed to be abandoned and open to occupation and entry by any person when it has remained unworked on working days for the space of seventy-two hours, unless sickness or other reasonable cause be shown to the satisfaction of the Gold Commissioner, or unless the grantee is absent on leave given by the Commissioner.

CHAPTER VIII.

THE ORIGIN OF GOLD PLACERS.

For those fragmentary deposits carrying gold known as "placers," we are indebted primarily to the great glaciers of the Ice Age, and after them, to the rivers of both past and present, and, in a more local and restricted sense, to the waves of the sea.

We might imagine what the mountains of the world would have been without these active agents of erosion. We may conceive of them as vast, smooth, rolling billows of strata, occasionally broken by stupendous cliffs, the result of profound faulting. As mineral veins are mainly due to the action of hot springs, geysers, fumaroles, etc., we might have noticed here and there along such lines of fault-fissure, mounds of tufa, of calcareous or siliceous matter, like those around the geysers of the Yellowstone, marking the position of the veins below in process of filling and formation.

THE WORK OF GLACIERS.

On such an uneroded country, let the glaciers be set to work, filling every fold and undulation of the surface. The ice sheets would plane off the tops of the mountains, exposing the rings of strata composing them, and the ice tongues, or glaciers, descending from the sheet, would cut deep, broad, U-shaped swathes down the sides of the hills, as shown in Fig. 1, and by mighty canyons expose the anatomy of the mountains and the veins in them. The débris from this planing would be distributed in windrows on the sides of the canyons and in moraines along ravines, to be winnowed and assorted by subsequent streams and rivers, and the finer material carried out onto and distributed over the plains, forming soil for the agriculturist, and a part to the ocean to form sea-bottoms.

ORIGIN OF DRIFT-GOLD.

If all the gold that has thus been spread far and wide in minute grains by these agencies could be collected, it would far exceed all that has been, is, or ever will be, obtained by man in his puny efforts at vein and placer mining and sand washing. We cannot suppose that all this gold, so widely distributed

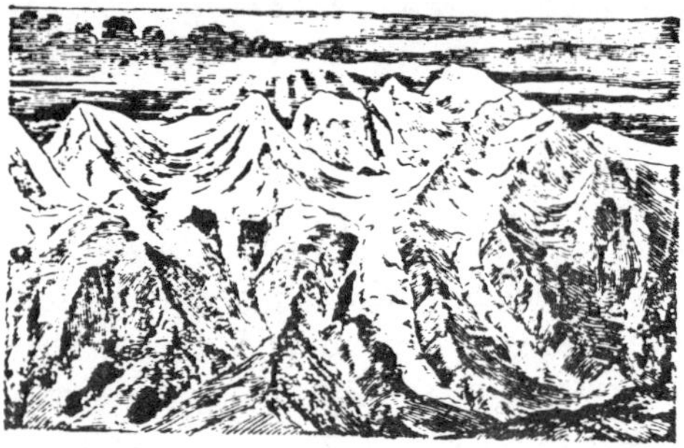

FIG. 1.

over the earth's surface, found more or less in every stream, and even among the waves of the seashore, and in regions far remote from any known gold veins or ore deposits, could all have come from well-defined, gold-bearing, quartz-fissure veins, but rather from the general breaking up of vast bodies and even mountains of crystalline rocks, such as granites, porphyries, lavas, and other igneous rocks containing more or less gold disseminated in minute particles throughout their masses.

ANCIENT PLACER FORMATIONS.

Though modern placer deposits are generally conceded to have been laid down by the action of comparatively recent glaciers, streams, and other bodies of water, in more or less loose, incoherent banks, yet there are other, far older formations, firmly consolidated into rock, which may be considered as ancient placers, having had the same alluvial origin as modern placers, and withal gold-bearing, such, for instance, as the uptilted gold-bearing conglomerates of the Transvaal (Fig. 2), and the gold-bearing Cambrian quartzites

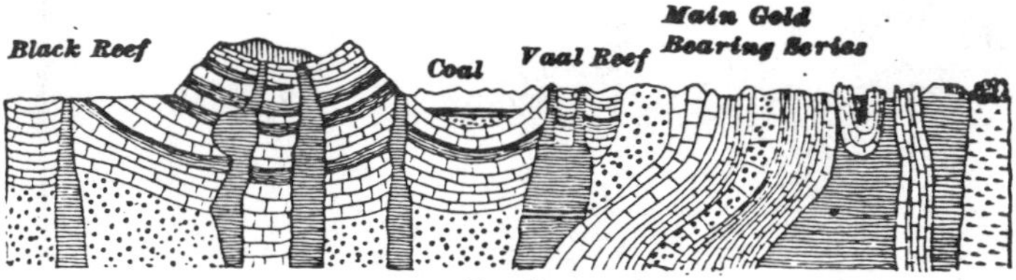

FIG. 2.

of the Black Hills of Dakota. These are but ancient placers, river beds, or sea beaches containing gold, consolidated by time into hard gold-bearing rock, and tilted up by the upheaval of the mountains.

REGIONS OF GLACIATION AND PLACER FORMATION.

Since glaciers are the parents of most of our large placer deposits, we must look for such deposits principally in those northern and mountainous regions which have been most subject to the reign of glaciers. Such, for instance, is Alaska, with its coast line torn to tatters by long glacial fiords, and its mountain ranges cloven through and through by the passage of ancient and modern glaciers, and the whole region, as well as that of the adjacent British Columbia, traversed by a network of streams and narrow lakes derived directly or indirectly from the glaciers. The same phenomena exist all through the Sierra region, down the Pacific coast to Southern California, and also in Idaho, Washington, and other northern regions. In all these, the placer deposits, both by glacier and stream, are often to be estimated by the hundreds of feet in thickness, while as we retreat inland towards the Rocky Mountain region, where precipitation was

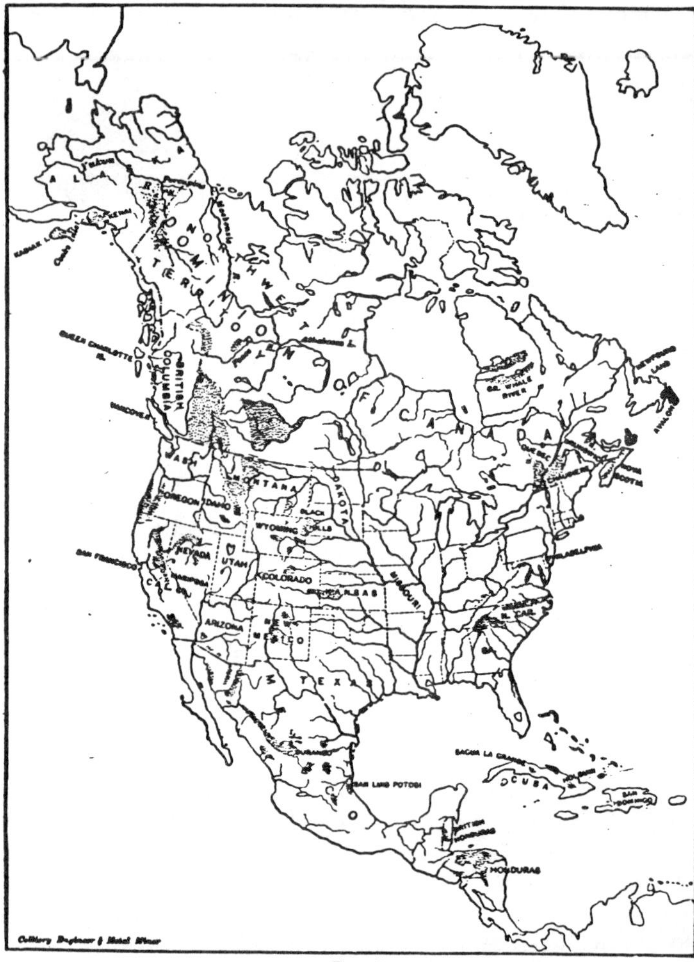

FIG. 3.

less, the deposits decrease in size and thickness. The placer areas are represented by the shaded portions of Fig. 3.

CHAPTER IX.

THE ORIGIN AND DEVELOPMENT OF PLACER MINING.

Gold washing of alluvial deposits, both in ancient and modern times, preceded vein, or lode, mining. In ancient times, vein mining seems to have been almost unknown; the gold of the ancients was entirely derived from the sands of streams. In modern times, while placer mining preceded, it often led to the discovery of veins and the developing of ore deposits in place. The discovery of gold at Coloma, near Sutter's Mill, California, in 1848, by James Marshall, is a well-known story. Marshall, while digging a race for a sawmill, found some pieces of yellow metal which he thought might be gold, a suspicion confirmed by one of his workmen, who had worked gold in Georgia. By the help of a "rocker," Marshall gleaned about an ounce of gold dust. (See Fig. 4.) This discovery led to examination of other California rivers having their sources in the Sierras, and soon every stream along the western slope of the Sierras was being worked for gold.

FIG. 4.

DEVELOPMENT OF PLACER MINING APPLIANCES.

At first only the crudest appliances were used, such as pick, shovel, pan, and rocker. Later the "Long Tom" sluice was introduced. Work on dry bars led to mining river bottoms by wing dams; then streams were turned from their natural courses by big flumes and ditches. From the shallow placers, the miners pushed back to the deep deposits, and worked them by Long Tom sluices. As the deep deposits of gravel were often poorer than the shallow placers, open cuts were necessary, and long sluices were found to run dirt faster than shovelers could supply it.

FIG. 5.

Edward Mattison, of Connecticut, thought he could dispense with the pick and shovel by using a stream of water under pressure to break down the banks of loose débris, and he conveyed a stream through a rawhide hose with a wooden nozzle, and discharged it against the bank, as shown in Fig. 5. The earth was torn from the bank and carried into sluices, dispensing with the labor of shoveling. This was the beginning of hydraulic mining.

GENERAL DESCRIPTION OF PLACER MINING.

The placer miner takes advantage of the natural forces that have been acting for ages. Frost, ice, mountain torrents, and the decay of rocks have broken down veins, liberated gold, and distributed it under gravel and sand in beds of both ancient and existing streams. These forces having done their work, the miner clears up and harvests what nature has mined for him. The operations of nature have been so vast, and so gigantic are the deposits, that pick, shovel, and pan are frequently inadequate, and more powerful appliances are required; thus, powder blows up the

FIG. 6.

deep, solid deposits, water under pressure undermines and washes away high banks of gravel, cranes and hoisting apparatus carry off the huge boulders, while for harder deposits stamp mills are requisite. Water is directed against a gold-bearing bank with the momentum acquired by falling from an elevation, or with the gentler force of a shorter fall if it runs down a sloping channel. The first is the hydraulic process; the second is sluicing. The first breaks up and disintegrates; the second assorts and concentrates. In hydraulic mining, the two are combined into one operation.

Water, falling through pipes from a height of one hundred or two hundred feet, is delivered through nozzles in continuous streams against a bank of earth, undermining it. (See Fig. 6.) The overhanging masses fall to the base, and are loosened and broken apart; the water penetrates every crack and pore; large boulders are thrown aside like pebbles; the whole mass is stirred and mingled, while the accumulated waters, thick with sand and earth, flow away down the slope, leaving the larger boulders and gold resting, clean washed, on the surface of the bed-rock. This process is applicable where the deposits above the lower gold-bearing stratum are so thick that they cannot be removed by digging. To do this there are required: First, a sufficient head or height and quantity of water; second, a rapid fall or slope from the base of the bank, so that water will flow swiftly away and carry the loosened gravel, sand, and earth with it.

ORIGIN OF FLUMES AND DITCHES.

In California and other gold-mining districts, high mountains give rise to numerous streams flowing towards and across the gold region, and the deep valleys and ravines permit of ample fall and drainage. These streams have to be diverted from their course and carried in ditches and flumes many miles, along easy grades. The best gold deposits are in trough-like basins, hemmed in by rocky walls, through which artificial outlets must be cut to get drainage. When the position, depth, and richness of a deposit are adapted to hydraulic mining, an outlet must be provided for the water. It may be there by nature, or it may involve the cutting of a tunnel through the "rim-rock" from an adjoining ravine to tap the lowest part of the basin and secure a vertical fall of from 50 to 100 feet to the base of the deposit. Many years and great sums of money have been spent on these undertakings. These tunnels may vary from a few hundred feet to a mile or more in length; their average dimensions are from six to eight feet wide by seven feet high.

In 1867, in California, there were 5,328 miles of artificial ditches, according to Blake, and 800 miles more of subsidiary ones, cut into the earth of the hillsides and crossing rocky points and deep valleys by flumes and pipes.

EXAMPLES OF DITCHES AND FLUMES.

There were ditches eight feet wide at the top, six feet wide at the bottom, and three feet deep, with a grade of from twelve feet to eighteen feet per mile. Large sheet-iron pipes were found better than wooden flumes and were generally adopted. One flume was 1,300 feet long, 260 feet above the surface, and supported on wooden towers. On the Truckee (or U) ditch there were thirteen miles of flume, eight feet wide and four feet deep, hung on the side of a deep canyon.

ORIGIN OF THE USE OF PIPES.

Flumes are difficult to keep in repair and are liable to be broken or blown down, and if left dry the boards warp and split, and the repairs of a flume cost 90 per cent. more than those of a ditch; hence, the substitution of wooden and iron pipes. Pipes, moreover, prevent loss of water by evaporation. Wooden pipes are made up of wooden staves bound together by round iron bands. Iron pipes are of stout sheet iron or boiler iron, and are made either in short joints, several of which are riveted together, forming sections about 20 feet long, or of continuous helical strips (the so-called "spiral-riveted" pipe) running the full length of a section. The thickness of the iron and the number and size of the rivets depend upon the pressure of the water; the smaller pipes are from 7 to 11 inches in diameter and are usually made of No. 20 iron. The sections are united on the ground and secured by wire wound around projecting eaves or hooks of iron upon each end of each section. The whole pipe is firmly fastened to the surface by posts set in the ground, to prevent it from rolling down slopes or being carried away by freshets or snowslides. In El Dorado County, California, a pipe is used to carry 50 inches of water across a valley 1,600 feet wide and 90 feet deep. The pipe is 10 inches in diameter, of iron $\frac{1}{16}$ inch thick; the supply is 90 feet higher than the delivery. Ditches are often built by companies, and the water sold to the miner at so much per miner's inch per day of 10 hours.

DISTRIBUTION OF GOLD IN DEPOSITS.

In a few cases, gold is rich in thin streaks of cemented gravel and alluvions on the top gravel. Even in high banks, the upper "top gravel," if of a fine, light, quartz wash, without boulders, has been washed at a profit; hence, the miner tests the whole deposit. A fine lamina is sometimes found at grass roots. Pay gravel may occur high above bed-rock, but generally the top gravel is not rich enough to pay, the gold concentrating in those strata within a few feet of bed-rock and on and in bed-rock itself. A section of a placer bed from surface to bed-rock is shown in Fig. 7.

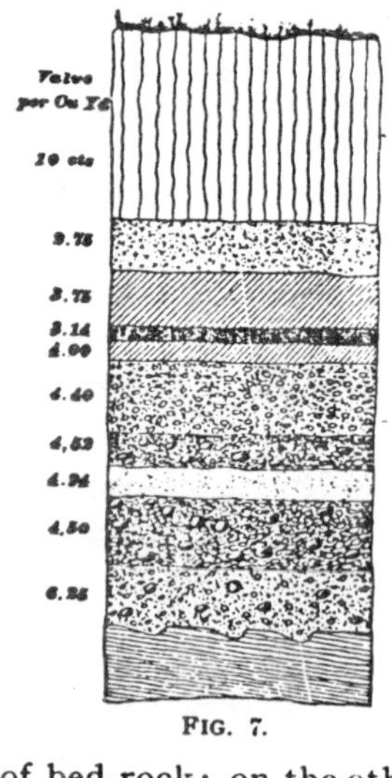

FIG. 7.

Sand is generally poorer than gravel or boulders. Rich pay may occur in undulations and depressions of bed-rock; on the other hand, deep holes caused by waterfalls are often, contrary to expectations, unproductive. Accumulations may assume the form of reclining cones, the apex resting on the top of the hillock, the gold being concentrated in the lower end of the deposit.

Placers in place, that is, lying in a gold vein, may contain deposits similar in quantity and distribution to those of the original vein on which they lie.

Placers of accumulation are the richest where the current of the stream was interrupted by diminution in its fall, by sudden change of direction, entrance of tributaries, or by reefs, bars, and eddies. (See (*a*), Fig. 8). Small depres-

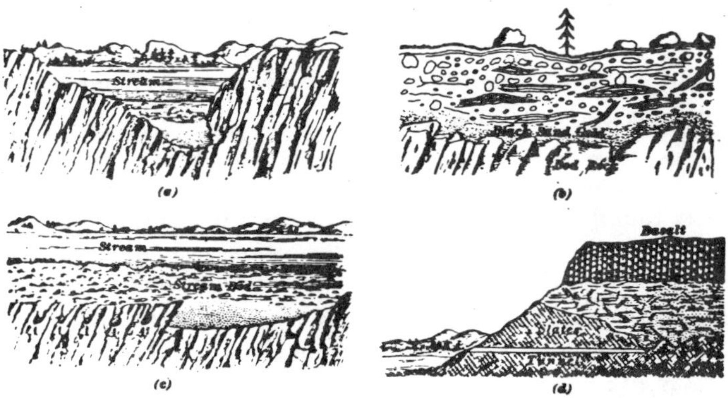

FIG. 8.

sions, creases, holes, and fissures (*A*, *A*) in bed-rock over which the current passes are likely to be rich. (See (*b*) and (*c*), Fig. 8.) As there are at times, in some placers, different periods of deposition, the lowest layers of each period are apt to be the richest. Several periods of deposition may have succeeded one another, and several rich strata occur on the same ground. Not only the courses of present streams, but also the channels of ancient rivers, are localities of placers; the latter are the so-called "deep leads." (See (*d*), Fig. 8.)

Alluvial layers may, at intervals, by a cementing process, form a seeming bed-rock, called a false bottom. There may be one or more such false bottoms before the true bed-rock is reached, and gold may occur on each of these.

Placers have generally been formed of material transported some distance by glaciers and drifts, but placer diggings sometimes occur on the very outcrop of decomposed gold veins, and are called "placers in loco," or in place. The deep leads were the work of a river system quite distinct from that now existing. Modern rivers, cutting across these old river courses, have redistributed their golden sands. Placers, as a rule, are much richer than the veins from which they have been derived.

Gold dust, under certain favorable chemical conditions, may amalgamate into masses, forming nuggets, such as are rarely found in veins. Shallow placers are often due entirely to the disintegration of quartz veins near by; in such cases, the drift will be barren above the point where the reef or vein crosses it. Plain diggings present a great variety, both in character and material of deposits.

Gold alluvions occur in river channels, in basins, and on flats, as surface deposits of sand and gravel, and as accumulations of clay, sand, gravel, pebbles, and boulders of all sizes, with, in some cases, caps of lava. Shallow placers are those whose deposits vary from a few inches to several feet, as distinct from deep placers, which cover large areas and are frequently several hundred feet in depth.

"Hill claims" are deposits of gravel on hills.

"Bench claims" are deposits occurring in bench-like forms on declivities above the line of existing rivers.

"Gulch diggings" are found in gulches and ravines.

"Flat deposits" occur in small plains, or "flats."

"Bar claims" are bars of sand and gravel on the sides of streams above water-level.

"Black sands" are the gold-bearing sands of the seashore.

"Surface mining" and "deep mining" are the two main divisions of placer mining. "Sluice," "drift," and "hydraulic digging" are local names.

EXAMPLES OF PLACER BANKS.

In the Ballarat gold fields of Victoria, the wash dirt is in a series of leads of varying widths, starting from the same point and trending in different directions towards the deep leads. In one placer, the width of the gutter and reef wash was 100 feet, and the depth of pay dirt 5 feet. The barren drifts overlying the pay dirt were of black clay, the reef of green slate, and the bottom of sandstone. At another place, wash dirt 6 feet thick was of a dark blue granite or green slate. At Melbourne, the formation is Upper Silurian schist, traversed by a metalliferous eruption of dikes. The gold lies in crevices in the rotten bed-rock, and in "potholes" in the ancient river bottoms. Nuggets have been found in soft clay and in the face of bed-rock.

CHAPTER X.

METHODS OF WORKING—SURFACE MINING; WORKING FROZEN GROUND: DRIFTING; HYDRAULICKING.

Surface mining is applied to operations in shallow placers, and in new districts frequently gives very large returns. These deposits are, however, limited in extent and soon worked out. In localities where water was scarce, dry washing was formerly resorted to. Selected rich dirt was pulverized and worked in a *batea*, or wooden dish, the earthy portions being separated by a circular motion given the dish. Gold was also extracted by winnowing.

Along the Pacific Coast, "beach mining" was practised. The gold occurs in the sands of the beaches in a very finely divided state, in layers of magnetic iron, or "black sand," where it has been concentrated by the action of wave and tide. By the wash of the waves, the lighter quartz sand is carried away, leaving the gold-bearing black sand behind, and this is then covered by barren material. When the tide ebbs, the miners scrape up the thin gold-bearing layers and transport the material to the washing places, where it is washed in sluices. These black sands assay from $10 to $30 per ton, only a part of which, however, can be recovered. Platinum occurs with the gold, in less flattened grains.

BAR AND RIVER MINING.

In the early days, river mining was extensively carried on. When the portion of the bars above water-level was exhausted, the miners extended their operations to those parts under water. This necessitated the streams being dammed or diverted into a new channel (see Fig. 9). Beds of rivers for long distances were laid bare, and much expense and risk were incurred from floods and breaking of dams; yet, though the losses were often great, the profits obtained in a short time were sometimes enormous. In some cases, instead of turning a river from its course, dredging machines have been used for the purpose of raising the material to the surface, where it is washed in the usual manner for its gold

FIG. 9.

contents. The Santa Fe Placer Mining Company is mining the beds of the Galisteo River and the Rio Grande by dredging. For this purpose, the Nettleton placer machine was designed. This machine is simply a powerful steam bucket dredge, with a capacity of one cubic yard of material per minute, having as an auxiliary a 6-inch centrifugal pump, whose suction pipe will extend down the dredge ladder to within 12 inches of the lowest point reached by the buckets. It is claimed that this pump will not only bring up the necessary water for sluicing, but also such loose material as may be left by the buckets, and will in a great measure clear the bed-rock of gold. The product of the dredge and pump is deposited in a sluice-box 25 feet above the deck of the barge, from which elevation the work is done by gravity until the material and water are disposed of. Passing down the first sluice of 30 feet, a grizzly or grating is reached, removing all stones over 3 inches in diameter, the finer material passing through screens which further reduce it to one-half inch and smaller. So the non-productive material is removed at once and deposited behind the barge. The percentage of fine flour gold being very large, the material is passed over a burlap sluice, the fibers of which arrest and hold the gold. The usual accompanying "black" or magnetic sand carries a great deal of gold, as much, at times, so it is reported, as $2,000 per ton. To save this, after passing the burlap the matter comes in contact with strong magnets placed in the circumference of a cylinder, the iron adhering to the magnets, from which it is removed by a revolving brush, the non-magnetic matter passing on to a revolving screen, where it is reduced to $\frac{1}{16}$ inch, preparatory to being run over amalgamated plates, such as are used in stamp mills, or into a series of boxes containing quicksilver. By this time all collectible gold will have been caught, and after being run out through traps to save any stray amalgam or quicksilver, the now barren material will pass into a tailings well, to be taken up by an 8-inch centrifugal pump and deposited far behind the boat. Fine sand settling in the riffles of the sluices or burlap may be treated with cyanide. Depressions in bed-rock which the dredge can not reach may be reached by an air caisson, and the bed-rock thoroughly cleaned.

Another plan suggested is to raise the material and water for sluicing, with a centrifugal pump, to the amalgamating plant placed on the bank. Large stones and gravel from the screens will be deposited in the excavation back of the workings; the fine tailings, and the sluice and surplus water will be conducted down the river by flume a sufficient distance to prevent their return. This plan will enable the bed-rock depressions and crevices to be cleaned by hand at less expense than by dredge or caissons. The water flow of the river will not exceed 10,000 gallons per minute during ten months in the year, so no great capacity of pumps will be needed. The natural conditions have made the Galisteo River a promising proposition. The extent of its gold deposits can only be conjectured.

Ground sluicing consists in treating gold-bearing gravel, dug out by pick and shovel, by washing it in trenches cut in bed-rock. It is like hydraulicking, except that the water is not used under pressure, and often wooden sluices are dispensed with entirely, the rough rock serving for riffles. The lighter material is carried away by the water, and the heavier dirt remaining behind is collected and worked by rockers. Ground sluicing is often adopted where there is not a sufficient amount of water for the constant use of a box sluice, and a head can only be gotten for a short period after heavy rains. A ground sluice is then used, if there is abundant fall and outlet for the tailings. It is a gutter worn by the water in its flow, the miner assisting the operation by loosening the earth with a pick. The pay dirt is washed in by the stream or conveyed thither by manual labor. If the bottom be a hard, uneven rock, its inequalities will suffice to arrest the gold; if not, a number of boulders, too heavy to be moved by the stream, are thrown carelessly into the sluice. This process saves only the coarse gold. To clean up, the water is diverted from the channel and the auriferous matter collected, to be panned out or cradled..

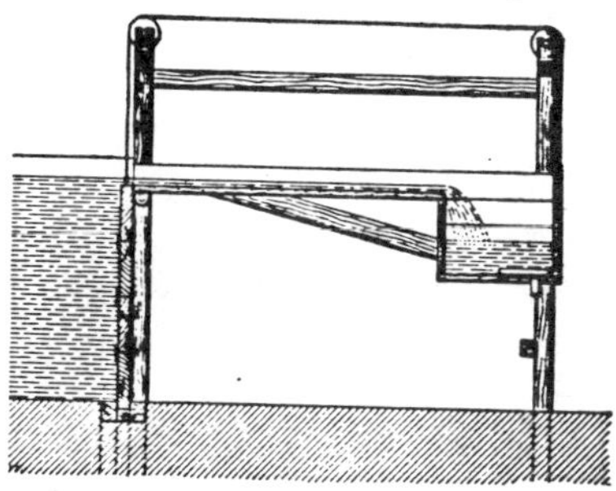

FIG. 1C.

BOOMING.

Booming is ground sluicing on a large scale, by means of an intermittent supply of water. The water is collected behind a dam with an automatic gate (Fig. 10), which, when the dam is full, opens, and the entire contents of the reservoir go down with a rush, carrying into the sluices all the material collected below. The rush of waters carries off boulders and dirt, leaving the heavier particles of gold and magnetic iron, or black sand, collected behind on the bed-rock floor.

WORKING FROZEN GROUND.

In the placer mines of Siberia and Alaska, the ground is frozen to a considerable depth. Frozen gravel will successfully resist all attacks of pick and shovel, and its extreme toughness renders even drilling and blasting very tedious and ineffective, so the miner thaws the ground before attempting to dig it. This he accomplishes by building a fire against the ground to be removed. In sinking his shaft, if the surface is frozen, he builds a fire of wood where he desires to sink, and the heat from this thaws out the ground for some little distance. The fire is rendered more effective by covering it with charcoal, thus confining the heat. When the fire dies down, the miner scrapes aside the embers and shovels away the loosened ground beneath, until he comes once more to the frozen portion, where another fire is built and the whole operation repeated; and thus down to bed-rock. The sides of the shaft are given what little support is necessary by a light cribbed timbering or a rough square set with lagging.

From the bottom of the shaft the miner starts his drift. He is obliged to thaw every foot of this, also. A strong wood fire is built against the face of the drift, covered with charcoal as before, and allowed to burn out; in other respects, the work is the same as in ordinary drifting. All workings must be tightly, though not heavily, timbered. This feature is frequently neglected, with many sad accidents as a result. Match-boards (tongued and grooved) are frequently used on the sides, though either round or split lagging will answer if closely set. In cross-cuts and chambers, the roof timbering should be entirely independent of that of the sides, as in Fig. 11. Large deposits are divided

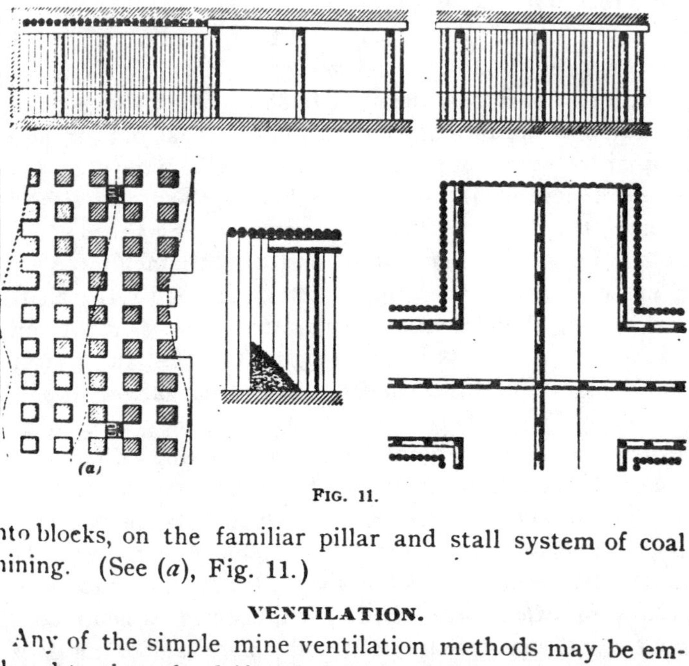

FIG. 11.

into blocks, on the familiar pillar and stall system of coal mining. (See (*a*), Fig. 11.)

VENTILATION.

Any of the simple mine ventilation methods may be employed to clear the drifts of the noxious gases generated by the fire, and to keep the air in the workings pure. Warm air and gas naturally rise to the roof of the drift and out of the shaft on the side nearest the drift, the cool air from the surface coming down the other side of the shaft and along the floor of the drift to the face. As the drifts get longer, particularly in small and low workings, the air from the face has more time to cool and diffuse with the fresh air, and the current gets weaker and weaker, so that it finally becomes necessary to use artificial aids to the ventilation. The simplest of these is by furnishing an artificial separation for the outgoing and incoming air-currents. A brattice is made by fastening canvas from top to bottom along the middle row of posts shown in Fig. 11, or boarding up between the posts with light plank. This is continued nearly to the face and usually up to the surface, dividing the workings into two passages, the cool, fresh air from above passing down one and forcing the warm air and gas out through the other to the surface. If the heat of the mine does not create sufficient draft, a fire can be built at the uptake side of the shaft, the draft created by the fire assisting the natural draft of the mine. Ventilation may also be facilitated by the use of a fan, either forcing fresh air down the shaft or exhausting the vitiated air; the latter form is better when the scale of the work is large enough to warrant the use of power-driven fans. For small workings, too narrow to permit division by brattices, a small hand fan may be used at the surface or at some point along the workings where the air is good, the air being carried to the face in tin or sheet-iron pipes or wooden boxes, and delivered towards one side and at the bottom of the face, this scheme giving the quickest and best ventilation. For long tunnels, a furnace system of ventilation is frequently used, the tunnel being connected with the surface by shafts, which are sunk or upraised at intervals as the work progresses. The hot air and gas from the fire at the face rise through the nearest shaft, and cool, fresh air replaces them, coming down the shafts further back along the drift. As soon as the face gets so far beyond the shaft that its ventilation becomes very slow and poor, a new connection is made with the surface; the use of a hand fan at the foot of the shaft or a light brattice carried along the side of the drift from the last shaft to the face, with a curtain behind the shaft to prevent a direct current of air between it and the previous shafts, and thus force the fresh air to follow the brattice, will allow the drift to be carried forwards considerably farther before again connecting with the surface.

The effect of the fires in the drifts is to raise the temperature to an oppressive point, so that, as in some of the large Siberian mines, the miners work stripped to the skin, though the temperature outside is many degrees below zero. An amount of wood equivalent to a thickness of one foot across the face will thaw out about the same depth of gravel. Fourteen inches, however, is about the maximum depth to which the thawing will extend.

DRIFTING.

Gold is mined in deep deposits by tunnels and drifts, notably in those districts where the deposits are covered by an overflow of lava, as in (*d*), Fig. 8. Drifting presupposes the concentration of the metal in a well-defined stratum or channel. When the existence of a pay channel is determined, it is opened up and developed by a tunnel run in such a manner as to drain all parts of the mine. The location of this tunnel is a matter of great importance. If the channel is discovered on the hillside, and rises as it enters the hill, the tunnel is run along its bed, following the bed-rock; otherwise, the tunnel is driven below the channel, or through the "rim-rock," in such a position that the lowest point of the deposit will be above it. Sometimes shafts are sunk and the gravel drifted out and raised through them to the surface. The tunnel once driven and the channel opened, drifts are run through the pay ground on both sides, and the material is breasted out, timbering being used as required. Gravel is removed from the tunnel in mine cars, to the mouth, where it is dumped on floors and washed in sluices. When too firmly cemented for simple washing, it is first crushed under stamps. In some mines, steam locomotives are used for transporting men and material through the tunnel, which may be more than a mile in length. In

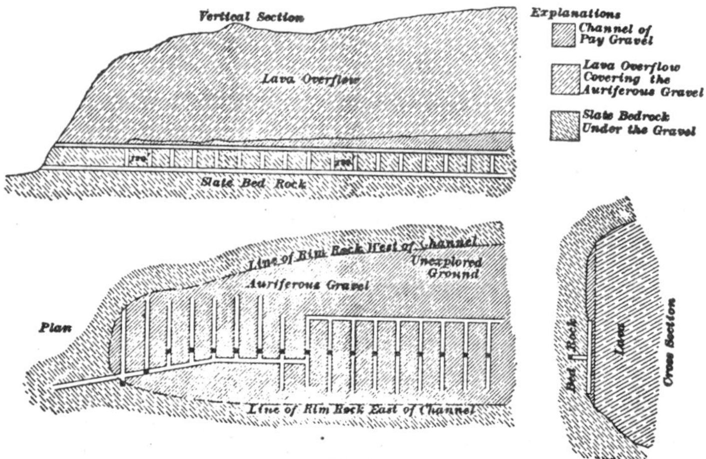

FIG. 12.

the "Sunny South" mine, Placer County, California, the main tunnel is below the channel, as shown in Fig. 12, thus allowing the mine to be worked conveniently.

HYDRAULIC MINING.

Hydraulic mining is that method in which the ground is excavated by water discharged against it under pressure. Deep placers, if sufficiently rich, may be worked by drifting, but hydraulicking is far the more economical. For hydraulicking there should be ample facilities for dump and grade, and a sufficient head and supply of cheap water. When the banks are too firmly cemented or are covered with lava, blasting becomes necessary to shatter them before water can be advantageously employed.

PRELIMINARY INVESTIGATION OF PLACERS.

The value of the gravel deposits is the first consideration. Its determination involves ascertaining the course of the channel, the depth and position of bed-rock, which may be under hundreds of feet of detritus, the size of the deposit, and an estimate of the yield of the ground and the cost of the work. The geology and topography of the deposit and its surroundings must be considered, to assist in determining the course of the channel, the depth of bed-rock, and the facilities for dump. The value of the gravel may be approximated by making shallow pits and washing the material obtained from them and from other available placers, as where the bank has been exposed in section by the cutting of a stream. A large enterprise requires preliminary prospecting by shafts down to bed-rock and by drifts. The water supply and facilities for dump should be carefully considered, and also the length of the working season.

Different colored gravels, red, rusty, and blue, are sometimes considered as good signs, but are not reliable. Black sand is often accompanied by gold, but may be barren.

EXAMPLES OF PROSPECTING.

The example of the Malakoff property illustrates the preliminary work which is necessary on large deep-placer enterprises. To determine the value of the claims and the feasibility of working them, four shafts were sunk to ascertain the position of the channel, the value of gravel, and the depth of bed-rock. The first shaft struck the bed-rock of the main tunnel at a depth of 207 feet; 135 feet of this was in blue gravel averaging 41 cents per cubic yard. From the bottom of this shaft, drifts were run on the course of the channel for a distance of 1,200 feet. The width of the channel was estimated at 500 feet. The total length of the explorations was over 2,000 feet. The average assay of the samples from the various drifts was $2.01 per cubic yard; the actual yield of over 21,000 tons was at the rate of $2.75 per cubic yard. The gross cost of the preliminary work, including the four shafts, was $66,956.20.

CHAPTER XI.

WATER SUPPLY—RESERVOIRS, DAMS, AND MEASUREMENT OF WATER.

RESERVOIRS—SOURCES OF WATER.

The water supply for placer operations is obtained from running streams, melting snows, and rains. The snow accumulates on the mountains during the winter, and the heavy rains and warm weather of the spring season cause rapid thawing of these snowbanks, and enormous volumes of water rush down the gullies and ravines. The placer miner impounds this surplus water in large storage reservoirs, for use during the dry season. In selecting a reservoir site the following points should be observed (Bowie):

1. A proper elevation.

2. The water supply from all creeks and springs and the catchment area.*

3. The amount of rain and snowfall.

4. The formation and character of the ground with reference to the amount of absorption and evaporation.

The elevation of a reservoir depends upon the location of the mines and the extent of the country which it is proposed to cover with a ditch. The reservoir should be located below the snow belt, if possible, and at the lowest point of the catchment area, in order to obtain the maximum supply of water therefrom. The average and minimum supply of water from all streams should be carefully determined. Rainfall is greater in mountain districts than in lower countries, and greatest on the slopes facing the direction

* Area draining into reservoir.

from which the moist winds blow. Snowfall measurements are taken on a level, and a given amount of snow is reduced to water, and the fall calculated as rain.

ABSORPTION AND EVAPORATION.

The most desirable formation of ground for a reservoir site is one of compact rock, like granite, gneiss, or slate. Porous rocks, like sandstone and limestone, are not so desirable, on account of their absorptive qualities. Steep, denuded slopes are best, as but little water escapes. The greatest slope gives the largest available quantity of water. Vegetation causes absorption. At the Bowman reservoir, in California, 75 per cent. of the total rainfall and snowfall (reduced to rain) is stored (Bowie).

A reservoir must be made large enough to hold a supply capable of meeting the maximum demands. The area of the reservoir should be determined, and a table made showing its contents for every foot of depth, so that the amount of water available can always be known. A longitudinal section through the center of the reservoir, with cross-sections and contour lines five feet above each other, virtually determines the height of the dam and the contents of the reservoir with the water at any depth. The Bowman reservoir contains about 1,050,000,000 cubic feet of water. The catchment area is 28.94 square miles. The cost of the reservoir and dams was $246,707.51. Besides the main reservoir, all mines have distributing reservoirs which receive the water from the main ditch for delivery to the underground claims. These are small and adapted only for a short run.

DAMS.

Dams are used for retaining the water in reservoirs, for diverting streams, and for storing in canyons débris coming from the mines.

FOUNDATIONS.

Foundations must be solid and water-tight, to prevent the settling of the dam, leakage under its base, and wear in front by water running over the top. Whenever possible, the foundation should be on solid rock. Gravel is better than earth, but requires sheet piling. Vegetable soil is unreliable; all porous matter, sand, and gravel must be stripped off until the solid ground is reached.

WOODEN DAMS.

Wooden dams are constructed of round or hewn logs one to two feet in diameter, laid in a series of cribs 8 to 10 feet square and pinned together by tree-nails. These cribs are filled in with loose rock. A layer of planking on the face of the dam makes it water-tight. Abutments may be constructed of stone, cement, or wooden cribs.

MASONRY DAMS.

The masonry of dams must be well laid in hydraulic cement, so that the stones cannot slide upon one another, and the dam as a whole so set that it cannot slide upon its base. Neither the material nor the foundation must be required to bear too great a pressure. The stones must *not* be laid in horizontal courses extending from front to rear. Binders should be used; there should be no continuous joints.

EARTHEN DAMS.

Earthen dams are used for reservoirs of moderate depth. They should be at least 10 feet wide on top; a height of over 60 feet is unusual. Fig. 13 shows a section of the Bowman dam in California, which is 100 feet high and 425 feet long.

WASTE DAMS.

A waste dam is a crib of round timbers 12 to 30 inches in diameter, notched and bolted together, and the whole fast-

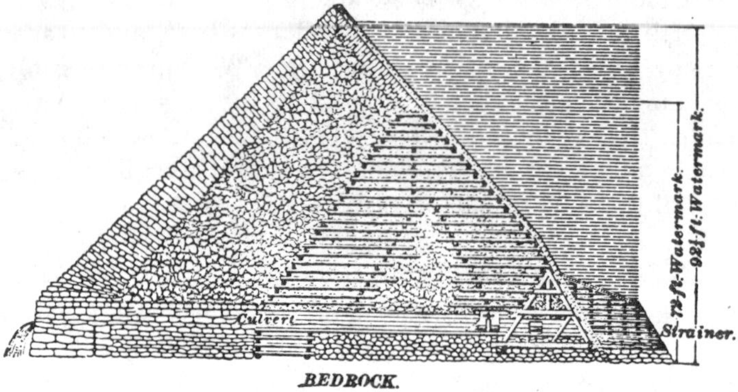

FIG. 18.

ened to bed-rock. The cribs are filled solid with rocks. The dam is provided with a number of waste-gates, each 40 to 50 square feet in area. These waste-ways are ordinarily kept closed, but are opened in times of freshets. The structure should be able to withstand any flood to which it is apt to be subjected, the waste water passing through the wastes and over the crest. Water passing over the dam falls onto bed-rock or onto a wooden apron.

DÉBRIS DAMS.

Débris dams are obstructions across the beds of streams to hold back tailings from the mines and prevent damage in valleys below. They may be stone, débris, wood, or brush.

MEASUREMENT OF FLOWING WATER.

Various forms of water meters are used for this purpose. Gauging by weirs of certain dimensions gives very close results. In this method the height of the surface of still water above the crest and some little distance back from the weir must be measured. There should be no considerable current to the water at the place of measurement. Flowing water is also measured by its discharge, under pressure, through orifices of regular section. The discharge of flumes of regular section may be calculated roughly from the mean surface velocity. An accurate calculation of the discharge of any stream may be made by multiplying the average velocity of the water at any point by the sectional area at that point. The discharge of small streams may be estimated by filling vessels of known capacity.

A right-angled **V** notch of thin sheet iron is a convenient form of aperture for measuring the discharge of water.

FIG. 14.

The discharge in cubic feet per second equals 0.0051 times the square root of the fifth power of the head, expressed in inches. The notch is fitted in one end of the weir box (see Fig. 14). The edge of the notch must be sharp and beveled off, and the inside face must be at right angles to the surface of the water. To keep quiet the surface of the water in the box, above the weir, baffle-boards are placed in the upper end of the box. The distance *a* of the surface of the water below the top of the weir is taken at a point some distance back from the notch (18 to 24 inches), where the water is level. This distance, subtracted from the total depth *H* of the weir, gives the head *h* of the water passing over the weir. The head is obtained as follows:

A straight-edge or level is placed on the weir plate P, extending back over the surface of the water in the box, and the distance a between its lower edge and the surface of the water measured. This distance subtracted from H (Fig. 14) leaves h. In gauging the quantity of water passing over a weir, the formula is

$$Q = 3.31\,L\,h^{\frac{3}{2}} + 0.007\,L.$$

Q is the quantity, or the discharge, in cubic feet per second; L the length of the weir; h the depth on the weir, corrected for velocity of approach; for this formula h must not be less than 0.07 feet.

In the accompanying table, opposite h, in the column Q, will be found the number of cubic feet of water flowing over the notch in one minute.

THE MINER'S INCH.

The "miner's inch" varies in every district, and is by no means a definite quantity, as the methods of delivering it differ in different places. It varies according to the head and the height of aperture. Usually, the head is 6 inches above the center of the aperture. The latter is a horizontal slit 1 inch high and 24 inches long, which can be closed so as to leave an opening of any desired length—1 inch long giving 1 inch of water, 10 or 12 inches long giving a corresponding amount of water; thus, in most districts, a miner's inch is considered as that quantity of water which will pass through an opening of 1 square inch area under a mean pressure, or head, of 6 inches. The quantity discharged from such an opening in 24 hours is equal to 2,274 cubic feet. A cubic foot is equal to 7.48 U. S. gallons, or 38 miner's inches. In some counties in California there are 10, 11, 12, and even 24-hour inches. Discharge apertures are rectangular, varying in width from 1 to 17 inches, and

DISCHARGE OF WATER THROUGH A RIGHT-ANGLED V NOTCH.

h Head, Inches.	*Q* Quant. Per Min., Cu. Ft.	*h* Head, Inches.	*Q* Quant. Per Min., Cu. Ft.	*h* Head, Inches.	*Q* Quant. Per Min., Cu. Ft.	*h* Head, Inches.	*Q* Quant. Per Min., Cu. Ft.	*h* Head, Inches.	*Q* Quant. Per Min., Cu. Ft.
1.05	0.3457	3.25	5.827	5.45	21.22	7.65	49.53	9.85	93.18
1.10	0.3884	3.30	6.054	5.50	21.71	7.70	50.34	9.90	94.37
1.15	0.4340	3.35	6.285	5.55	22.20	7.75	51.16	9.95	95.56
1.20	0.4827	3.40	6.523	5.60	22.70	7.80	51.99	10.00	96.77
1.25	0.5345	3.45	6.765	5.65	23.22	7.85	52.83	10.05	97.98
1.30	0.5896	3.50	7.012	5.70	23.74	7.90	53.67	10.10	99.20
1.35	0.6480	3.55	7.266	5.75	24.26	7.95	54.53	10.15	100.43
1.40	0.7096	3.60	7.524	5.80	24.79	8.00	55.39	10.20	101.67
1.45	0.7747	3.65	7.788	5.85	25.33	8.05	56.26	10.25	102.92
1.50	0.8432	3.70	8.058	5.90	25.87	8.10	57.14	10.30	104.18
1.55	0.9153	3.75	8.332	5.95	26.42	8.15	58.03	10.35	105.45
1.60	0.9909	3.80	8.613	6.00	26.98	8.20	58.92	10.40	106.73
1.65	1.0700	3.85	8.899	6.05	27.55	8.25	59.82	10.45	108.02
1.70	1.1530	3.90	9.191	6.10	28.12	8.30	60.73	10.50	109.31
1.75	1.2400	3.95	9.489	6.15	28.70	8.35	61.65	10.55	110.62
1.80	1.3300	4.00	9.792	6.20	29.28	8.40	62.58	10.60	111.94
1.85	1.4240	4.05	10.100	6.25	29.88	8.45	63.51	10.65	113.26
1.90	1.5220	4.10	10.410	6.30	30.48	8.50	64.45	10.70	114.60
1.95	1.6250	4.15	10.730	6.35	31.09	8.55	65.41	10.75	115.94
2.00	1.7310	4.20	11.060	6.40	31.71	8.60	66.37	10.80	117.29
2.05	1.8410	4.25	11.390	6.45	32.33	8.65	67.34	10.85	118.65
2.10	1.9550	4.30	11.730	6.50	32.96	8.70	68.32	10.90	120.02
2.15	2.0740	4.35	12.070	6.55	33.60	8.75	69.30	10.95	121.41
2.20	2.1960	4.40	12.420	6.60	34.24	8.80	70.30	11.00	122.81
2.25	2.3230	4.45	12.780	6.65	34.89	8.85	71.30	11.05	124.21
2.30	2.4550	4.50	13.140	6.70	35.56	8.90	72.31	11.10	125.61
2.35	2.5900	4.55	13.510	6.75	36.23	8.95	73.33	11.15	127.03
2.40	2.7300	4.60	13.890	6.80	36.89	9.00	74.36	11.20	128.45
2.45	2.8750	4.65	14.270	6.85	37.58	9.05	75.40	11.25	129.90
2.50	3.0240	4.70	14.650	6.90	38.27	9.10	76.44	11.30	131.35
2.55	3.1770	4.75	15.040	6.95	38.96	9.15	77.49	11.35	132.81
2.60	3.3350	4.80	15.440	7.00	39.67	9.20	78.55	11.40	134.27
2.65	3.4980	4.85	15.850	7.05	40.38	9.25	79.63	11.45	135.75
2.70	3.6660	4.90	16.260	7.10	41.10	9.30	80.71	11.50	137.23
2.75	3.8380	4.95	16.680	7.15	41.83	9.35	81.80	11.55	138.73
2.80	4.0140	5.00	17.110	7.20	42.56	9.40	82.90	11.60	140.23
2.85	4.1960	5.05	17.540	7.25	43.30	9.45	84.01	11.65	141.75
2.90	4.3820	5.10	17.970	7.30	44.06	9.50	85.12	11.70	143.28
2.95	4.5740	5.15	18.420	7.35	44.82	9.55	86.24	11.75	144.82
3.00	4.7700	5.20	18.870	7.40	45.58	9.60	87.37	11.80	146.36
3.05	4.9710	5.25	19.320	7.45	46.36	9.65	88.52	11.85	147.91
3.10	5.1780	5.30	19.790	7.50	47.14	9.70	89.67	11.90	149.48
3.15	5.3880	5.35	20.260	7.55	47.92	9.75	90.83	11.95	151.05
3.20	5.6050	5.40	20.730	7.60	48.72	9.80	92.00	12.00	152.64

1 cubic foot contains 7.48 U. S. gallons; 1 U. S. gallon weighs 8.34 pounds.

in length from a few inches to several feet. The discharge may be through 1-inch, 1½-inch, 2-inch, or 3-inch planks with square edges. The bottom of the opening may be either flush with the bottom of the box or raised above it. The head may denote the distance above the center of the aperture or above its top, and varies from 4½ inches to 12 inches above the center of the aperture (see Fig. 15). An aperture 12 inches high by 12¾ inches wide, through a 1½-inch plank, with a head of 6 inches above the top of the opening, gives a discharge of 200 miner's inches.

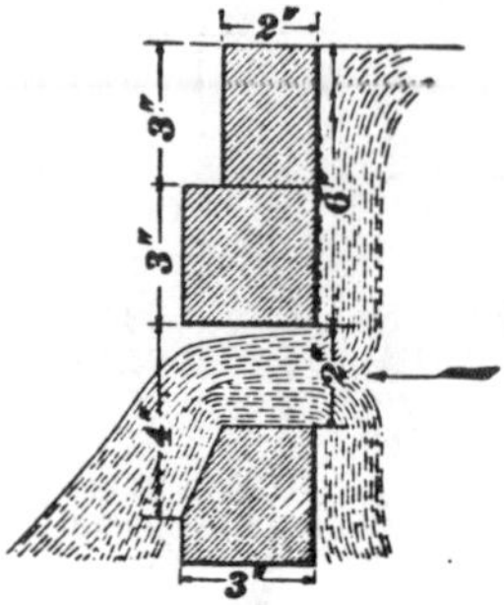

FIG. 15.

CHAPTER XII.

WATER SUPPLY—DITCHES AND FLUMES.

Thousands of miles of ditches have been made in the placer-mining districts of this country. On account of the rocky character of the country in such districts, steep grades are necessary, and high trestles with flumes, and wrought-iron and wooden pipes were built for carrying water across canyons and ravines. In constructing ditches, the following rules should be observed (Bowie):

1. The source of supply should be at sufficient elevation to cover the greatest range of mining ground at the smallest expense, great hydrostatic pressure being desirable.

2. An abundant and permanent supply of water should be assured during the summer months.

3. The snow-line should be avoided, and the ditch in snow regions located so as to have a southern exposure.

4. All the watercourses on the line of ditch should be secured, their supply counteracting the loss by evaporation, leakage, and absorption.

5. At proper intervals waste-gates should be arranged to discharge the water without danger to the ditch.

6. Ditches are preferable to flumes.

DITCHES.

SURVEYING A DITCH LINE.

Careful barometrical observations should be made to approximately determine the elevations, not only of the termini, but of intermediate points, from which surveying parties can start on the subsequent location of the line.

These points established, the line is surveyed and started. In leveling, turning points should be made on grade, the stations numbered and staked, and the pegs driven to grade. Every four or five stations the rodman calls off the reading, which is checked from the notes of the instrument man. Stations may be from 50 to 100 feet apart. Bench-marks should be placed every half mile. All details of tunnels, cuts, and depressions requiring fluming or piping should be worked out in full; a hand-level can be used for this purpose. Complete notes should be made regarding the ground along the center line.

The size of a ditch is regulated by its requirements. The smallest section for any given discharge is when the "hydraulic depth" is one-half of the actual depth. The hydraulic depth is the quotient obtained by dividing the area of the cross-section of the stream at any point by the wetted perimeter at that point. Trapezoidal and rectangular forms are adopted for ditches and flumes, respectively. The resistance due to friction in the latter form is smallest when the width is twice the height. Half a regular hexagon is a common form for ditches. In a mountainous country, with rocky soil, narrow and deep ditches with steep grades are adopted in preference to wider ditches with gentler slopes, as they are cheaper to excavate and repair. Ditches with grades of 16 to 20 feet per mile are quite common. Before commencing work, the line must be cleared of trees and brush; on the flume line, the brush for at least 10 feet on each side is burned off. On a hillside, the lines

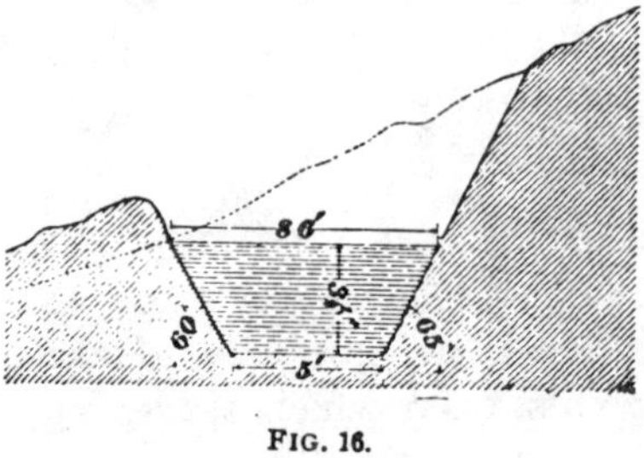

FIG. 16.

should be graded off so that the ditch may have walls of solid, untouched ground, and not made banks. Banks should be at least 3 feet wide on top. The slope of large ditches for mountain regions is usually 60 degrees for the upper and 50 degrees for the lower bank, but varies with the nature of the ground. The cost of digging is estimated at so much per cubic yard. The annual expense of running and maintaining large ditches averages about $400 per mile. The North Bloomfield ditch, shown in section in Fig. 16, is 55 miles long, 8.65 feet wide on top, 5 feet wide at the bottom, and 3½ feet deep; cost, $466,707. Its grade is 16 feet per mile; discharge, 3,200 miner's inches.

FLUMES.

Flumes are to be avoided if possible, being liable to decay and a continual source of expense. Instances occur, however, necessitating them, as where water must be carried along the face of vertical cliffs. Flumes usually have a slope of from 25 to 35 feet per mile, and are consequently proportionately smaller than ditches. They are usually made of seasoned pine planks 1½ to 2 inches thick, 12 to 24 inches wide, and 12 to 16 feet long. The edge joints are battened on the inside with pine strips 3 to 4 inches wide and ½ inch thick. The structure is reenforced every 4 feet by a framing consisting of a sill, cap, and two posts. A flume 4 feet wide by 3 feet high requires 4-in. × 5-in. posts and caps and 4-in. × 6-in. sills, with 8-in. × 10 in. stringers. The posts are set into the sills with a gain of 1¼ inches and not mortised. Sills should

FIG. 17.

extend 12 to 20 inches beyond the posts, which should be braced. Flumes should be built on solid beds or rigid trestles, as in Fig. 17. In carrying a flume around a hillside, the bed should be graded out and the flume placed close into the bank, to avoid danger from snowslides, etc. Curves should be laid with care, to ensure the maximum flow of water. The boxes must be cut in two, three, or four parts, necessitating more sills, posts, and caps. For good curving, the side planks are sawed partially through in places, so as to bend easily. To distribute water equally over an entire flume and prevent slack water, irregular curves, and splashing, the outer side of the flume is raised in accordance with the degree of curve and grade. Wastegates should be placed every half mile. In the snow belt, flumes are covered with sheds in places exposed to snowslides. Placing the flume close into the bank lessens the danger of freezing. If anchor ice forms on the bottom, the water should be turned out. Snow can be gotten rid of in the same way.

DETAILS OF CONSTRUCTION.

The bed being prepared, the stringers are laid and the sills placed upon them 4 feet apart. Bottom planks are nailed to the sills, the end joints being carefully fitted. The side planks are nailed to the bottom planks and posts, which are set in gains in the sills, an occasional cap being placed on the posts to hold the flume in shape. Sixteen and twenty-penny nails are used. The joints are then battened. Each box, when completed, is set on grade and wedged.

Where a flume connects with a ditch, the posts, for a distance of several boxes back, are lengthened, to permit the introduction of an additional plank on each side. The end boxes of the flume are flared, to permit a free entrance and discharge of water. At the junction with the ditch or in passing through a bank of earth, an outer siding is nailed to the posts to protect the flume. Exact sizes of lumber should be prepared at the mills, so that the flume can be rapidly constructed. Enough water is turned into the flume as the work progresses to float down the timber. The flume may in places be trestled with supports every 8 to 12 feet. The life of a flume will not exceed 20 years at most, and generally 10 years.

BRACKET FLUMES.

Flumes are frequently carried along precipitous canyons and cliffs on brackets fastened to the face of the cliff. In

FIG. 18.

Butte County, California, a line of ditch is run 200 yards up the canyon along a perpendicular wall of basalt. For a distance of nearly 500 feet the flume is carried on brackets along the face of the cliff, 118 feet above the bed of the ravine at its deepest point, and 232 feet below the top. The method of hanging is shown in Fig. 18. The brackets are made of 30-pound T rails bent into the form of an L; the longer arm—10 feet long—on which the bed of the flume rests, is placed horizontally, having its end supported in a hole drilled in the rock. The short arm—2 feet long—stands vertically, and has in its upper end an eye into which is hooked one end of a ¾-inch round iron rod, connecting to a ring bolt soldered into a hole drilled in the face of the cliff. Brackets are set 8 feet apart and tested to stand a weight of 14½ tons. The flume is 4 feet wide and 3 feet deep, with a capacity of 3,000 miner's inches. The figure shows a trestle 86 feet high. Along the line of the ditch is a trestle 1,080 feet long and 80 feet high. Another has been built 136 feet high. The total length of ditch and flume is 33⅓ miles.

CHAPTER XIII.

WATER SUPPLY—PIPES AND NOZZLES.

WOODEN PIPES.

For moderate heads, wooden-stave pipes are coming into use. They are practicable for any desired head, but are only economical to the point where the pressure necessitates such close banding that the cost exceeds that of iron or steel pipes of the same length. If kept full of water, the staves will last indefinitely, and the bands may be protected from rust by a coating of asphalt or other mineral paint. The amount of iron in the bands for each foot of pipe is the same as that in a foot of sheet-iron pipe of the same diameter, calculated to withstand the same head of pressure, with a considerable margin of safety.

IRON PIPES.

Wrought-iron or steel pipes are used exclusively for very high heads. For lower heads, either wood or iron may be used, the selection between them being a matter of location and cost. Pipes are used as water conduits, replacing ditches and flumes; as supply or feed pipes, conveying water from the pressure boxes to the claim; and as distributing pipes, taking water from the distributors or gates at the end of the supply pipe and delivering it to the discharge pipe or nozzle, which is usually made of sheet iron. Pipes used for conveying water across depressions are called *inverted siphons*. The thickness of metal for iron pipes is determined by the pressure of the water and the diameter of the pipe. Pipe once put together soon becomes water-tight from the foreign matter in the water. This result may be hastened by throwing in a few bags of sawdust. Pipes thus prepared will remain tight when subjected to a pressure of over 200 pounds per square inch. The Texas pipe, Nevada County, California, is an inverted siphon, 4,438.7 feet long and 17 inches in diameter, of riveted sheet iron. Its inlet is 304 feet above the outlet, and with full head its discharge is 1,260 miner's inches. The maximum head is 770 feet, equivalent to a pressure of 334 pounds per square inch.

JOINTS.

Pipes in general are 11, 15, 22, 30, and 40 inches in diameter, of riveted sheet iron or steel, Nos. 8, 10, 12, 14, or 16 (Birmingham gauge), made in sections of 30 to 36 inches, and riveted into lengths of 20 to 30 feet. The latter are put together stove-pipe fashion, neither rivets, wire, nor other contrivances being necessary. Where there is great pressure, iron collars or lead joints are used.

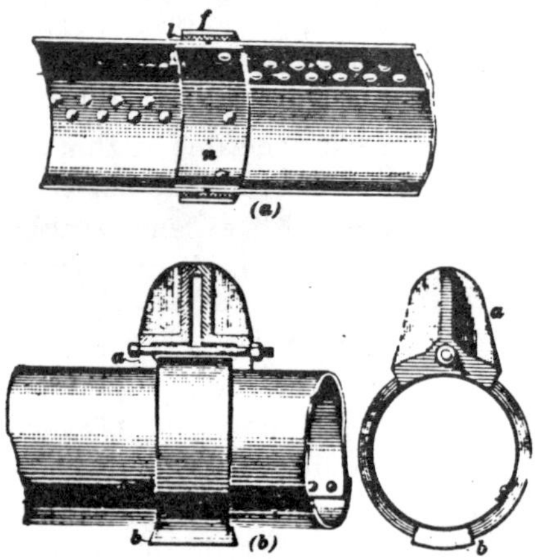

FIG. 19.

Fig. 19 (*a*) shows a style of joint which is frequently used. *f* is a wrought-iron collar 5 inches wide, $\frac{1}{10}$ inch thicker than the pipe-iron and with a play of $\frac{3}{8}$ inch between the inside of the collar and the outside of the pipe; *l* is the lead, which is run in and calked tight from both sides; *n* is a nipple of No. 9 iron, 6 inches in width, riveted in one end of each section by $\frac{3}{8}$-inch rivets. Fig. 19 (*b*) shows the method of tightening leaky joints. *a* shows the clamp and its method of application for forcing back the lead which has worked out by the expansion and contraction of the pipe. This is shown both in perspective and cross-section. The clamp *b* is used to keep the lead in place after it has been forced back by the clamp *a*. The two lower views of this clamp show the side and end elevations. Fig. 20 shows the elbow used in making short curves. *a, a* are angle irons riveted to the elbow on the outside of the curve and connected by iron straps with the corresponding angle irons in the pipe, as shown.

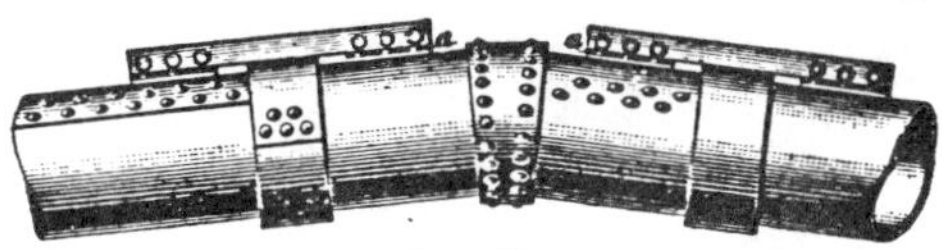

FIG. 20.

AIR VALVES; BLOW-OFFS.

To allow the escape of air from the pipe while filling, and also, in case of a break, to prevent the formation of a vacuum and the collapse of the pipe, blow-offs or air valves are provided. The simplest form is a loaded flap valve of leather on the inside of the pipe, arranged to cover an opening 1 inch to 4 inches in diameter. Another simple automatic valve is shown in Fig. 21. This sinks and opens when the water leaves it, and shuts when the water rises to it. Fig. 22 shows a form of blow-off valve used in

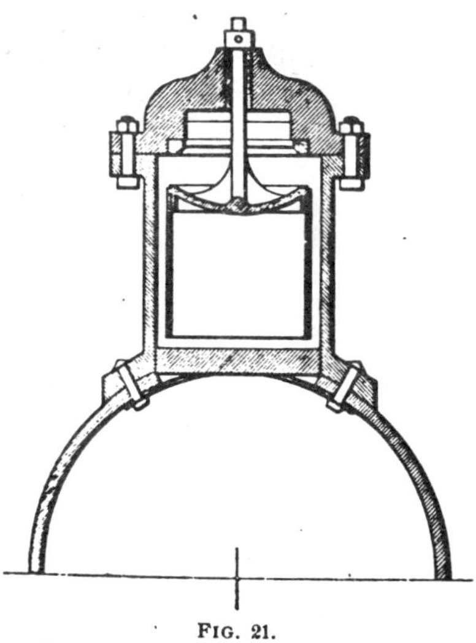

FIG. 21.

low places along the pipe line. Fig. 23 shows another automatic water-tight vacuum valve which is used at the high points on the line. The valve on the right is kept closed while the pipe is full, being opened occasionally to blow off air which may accumulate. The main valve is opened and the pressure of the water keeps the automatic valve closed. In event of a break in the pipe at a point beyond this valve, the pressure on the inside of the pipe is released, the automatic valve falls and admits the air, preventing a vacuum. On refilling the pipe, this valve, being open, allows the air to escape, closing only when the water reaches it.

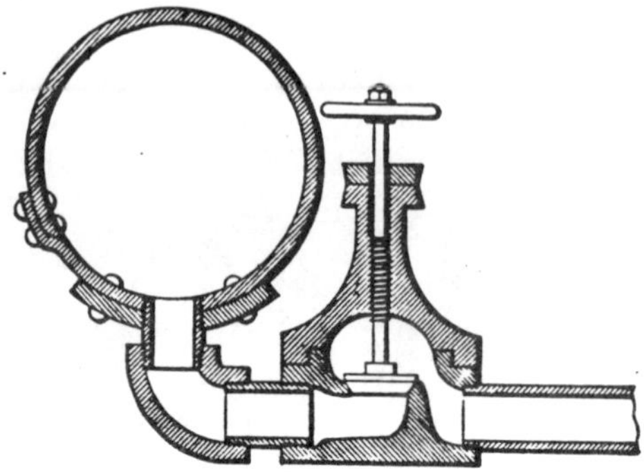

FIG. 22.

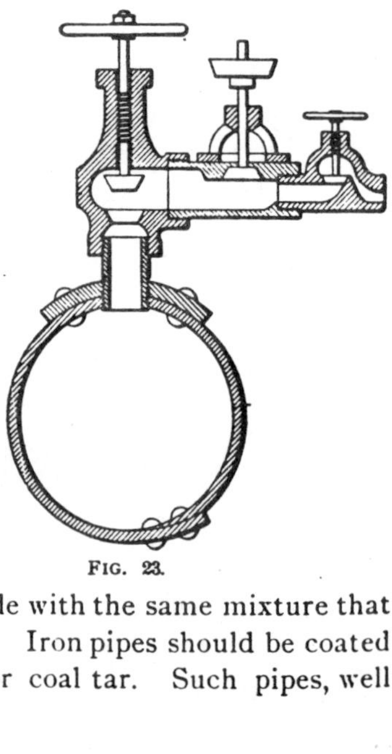

FIG. 23.

LAYING PIPES.

To preserve the pipe, it should be laid in a trench and covered with earth to a depth of at least a foot. Wooden pipes should be painted on the outside with the same mixture that is used for covering the bands. Iron pipes should be coated inside and out with asphalt or coal tar. Such pipes, well coated, are still in good condition after 15 years of service. The following mixtures are found to give the best results:

Crude asphalt.................28 per cent.
Coal tar (free from oily matter)...72 per cent.
Or, Refined asphalt...............16.5 per cent.
Coal tar (free from oily matter)...83.5 per cent.

To prepare these, asphalt in small pieces and coal tar are heated to about 400° F. and well stirred. The pipe is dried and immersed in the mixture, where it remains until it acquires the same temperature as the bath. When coated it is removed and placed on trestles, to drip and dry in the sun and air. For convenience of immersion, wrought-iron troughs 30 feet long by 3 feet wide and 2 feet deep are used. No. 14 iron requires 7 minutes' immersion; No. 6, 12 to 15 minutes'.

FILLING PIPES.

Pipes should be filled in such a manner as to prevent, as far as possible, the admission of air, which will be drawn into the pipes along with the water in surprising quantities, unless considerable care is taken. The best plan is to put a gate in the pipe below the intake, and thus regulate the flow, maintaining a steady pressure. A common form of penstock or sand box for intakes is shown in Fig. 24. A grating of bars should be provided to catch all drift. The water at the intake of the pipe should be kept quiet and sufficiently deep to prevent any air from being carried into the pipe. For this purpose the box is sometimes divided into compartments, one of which receives the

FIG. 24.

water and discharges it quietly into the second through lateral openings. There should be no difference between the water supply and the discharge. Some pressure boxes are arranged for two pipes.

SUPPLY OR FEED PIPES.

The water is conveyed in iron feed-pipes from the pressure boxes to the claim, and distributed to the discharge pipes by means of iron gates. The supply pipe is funnel shaped where it connects with the pressure box, and from there on it is usually of uniform diameter to the gate or discharge nozzle. Where 22 to 30-inch pipes are used, lighter iron than No. 14 B. G. is not advisable. The main supply pipe should descend in the most convenient and direct line into the diggings, avoiding angles, rises, and depressions. Air valves should be arranged at proper distances to allow the escape of air when filling the pipe, and prevent collapse. The pipe is braced and weighted at all angles. In filling the supply pipe, the water should be turned on gradually. Leakage in slip joints can be stopped with sawdust. Wherever a junction is to be made with another line, or the stream divided, the present practice is to fork the main pipe, cast-iron gates being placed in each branch. Fig. 25 shows the form of gate generally used.

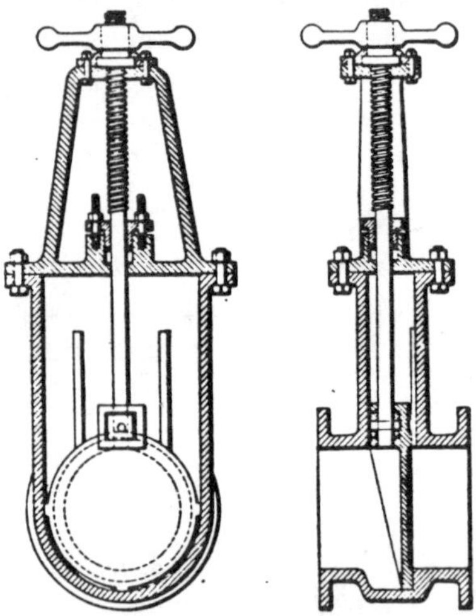

FIG. 25.

EVOLUTION OF THE GIANT.

We have mentioned how hydraulicking began with the use of a rude hose to break down the banks of débris, and so dispense with pick and shovel.

GOOSE-NECK.

The first improvement on this primitive device was a flexible iron joint formed by two elbows, one above the other, with a coupling joint between them, shown in (*a*), Fig. 26. These elbows were called "goose-necks." They were defective in design. The pressure of water caused the joints to move hard, and when the pipe was turned it would "buck" or fly back, endangering the life and limbs of the operator.

GLOBE MONITOR.

The goose-neck was succeeded by the "Craig Globe Monitor," which is shown in (*b*), Fig. 26. This was a simple ball-and-socket arrangement, but was very difficult of manipulation.

HYDRAULIC CHIEF.

The invention of the "Hydraulic Chief," by F. H. Fisher, was the next step. The machine is shown in (*c*), Fig. 26. The main improvements consisted of two elbows, placed in reversed position when in right line, connected by a ring in which there were anti-friction rolls. The ring was bolted to a flange in the lower elbow, but allowed the upper a free horizontal movement, while the vertical motion was obtained by means of a ball-and-socket joint in the outlet of the top elbow. The interior was unobstructed by bolt or fastenings, and the man at the pipe could operate it by means of a lever, without personal danger. Vanes, or rifles, were inserted in the discharge pipe to prevent the rotary movement of the water caused by the elbows, and to force it to issue in a solid stream. These machines soon became leaky.

DICTATOR.

The "Hoskins Dictator," the next step of the series, was a one-jointed machine with an elastic packing in the joint, instead of two metal faces. The joint worked up and down on pivots, and in rotating it the wheels ran around against the flange.

LITTLE GIANT.

The "Little Giant," a subsequent invention of Mr. Hoskins, on account of its simplicity and durability, super-

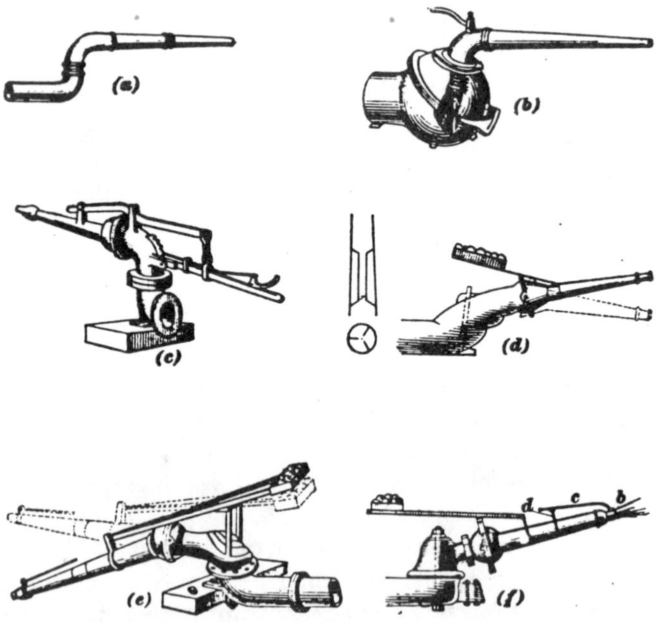

FIG. 26.

seded all previous machines. (See (*d*), Fig. 26.) It is a two-jointed machine, portable and easily handled, having a knuckle joint and lateral movement. The Giants have rifles, and nozzles from 4 to 9 inches in diameter, 5½ to 7-inch nozzles being commonly used. To keep Giants from bucking, they must be firmly bolted to a heavy piece of timber securely braced against gravel or rock. The machine and adjacent length of pipe must also be weighted to the ground. The bearings should be lubricated with tallow or axle grease.

HYDRAULIC GIANT.

The "Hydraulic Giant" (see (*e*), Fig. 26) is a modification of the Little Giant. The "Monitor," with a deflecting nozzle, invented by H. C. Perkins, is shown in (*f*), Fig. 26. The deflecting nozzle *b* permits the direction of the stream at any desired angle. When the lever *c* is in the rest *d*, the deflecting nozzle *b*, being of larger diameter than the nozzle, allows the stream of water from the nozzle to pass through without obstruction. To move the pipe, the lever *c* is taken from the rest and thrust in the direction in which it is desired to throw the stream. Any movement of the lever *c*, either to the right or left or up or down, throws the end of the nozzle *b* into the stream of water. The force of the water striking *b* causes the entire machine to swing around in the desired direction. Hoskins' deflecting nozzle is of cast iron, the same size as the main nozzle, to which it is attached by a packed universal joint. The operation is similar to that of the Monitor deflector. There is the disadvantage of a constant interference with the stream of water, and this nozzle is, therefore, somewhat dangerous.

CHAPTER XIV.

PLACER MINING PRACTICE—DEVELOPMENT OF GOLD-WASHING APPARATUS.

THE PAN.

The gold miner's pan, shown in Fig. 27, is pressed from a single sheet of Russia iron. It is usually about 12 inches in diameter at the bottom and 3 to 4 inches deep, the sides being inclined at an angle of 30 degrees from the horizontal and turned over a wire around the edge to strengthen the rim. It is used in prospecting, cleaning gold-bearing sand, collecting amalgam in the sluices, and throughout the business generally. Its manipulation requires skill and practice. A quantity of the dirt to be washed is placed in the pan, occupying about two-thirds of its capacity; the pan with its contents is immersed in water and the mass stirred, so that every particle may become soaked. When the dirt is soaked, the pan is taken in both hands, one on each side, and without allowing it to entirely emerge from the water it is suspended in the hands, not

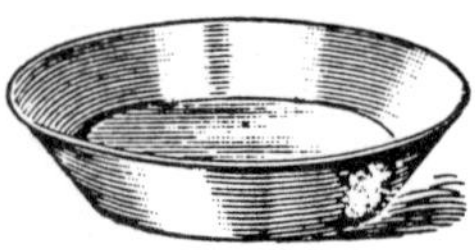

FIG. 27.

FIG. 28.

quite level, but tipping slightly away from the operator. In this position it is shaken so as to allow the water to disengage all the light, earthy particles and carry them away. (See Fig. 28.) This done, there will remain varying proportions of gold dust, heavy sand, lumps of clay, and gravel stones; these last are thrown out. A turn of the wrist allows the muddy water to escape in driblets over the depressed edge of the pan, without exercising so much force as to send the lighter portions of the gold after it. At last nothing remains in the pan but gold dust and heavy black sand and earthy matter. By the final careful working, with plenty of clear water, the earthy matter can be completely removed, but the heavy iron sand can not be gotten rid of by any method based upon its specific gravity as related to that of gold. If this iron sand be magnetic, the grains can be removed by a magnet. If there are fine particles of pyrites in the pan, they can generally be distinguished from gold by their lighter color, the gold being commonly a rich orange color.

THE BATEA.

The batea is a modification of the pan, used principally in parts of Mexico and South America where water is scarce. It is a shallow, conical bowl, turned out of a single piece of hard wood, and is about 20 inches in diameter by 2½ inches deep in the center. It is, on the whole, a better instrument than the pan for gold washing, the wood surface facilitating the concentration of the gold.

PUDDLING BOX.

The puddling box is a wooden box about 6 feet square by 18 inches deep, arranged with plugs for discharging the contents. The box is filled with water and gold-bearing clay. By continuous stirring with a rake, the clay is broken up in the water and run off. The concentrated material in the bottom is subsequently washed in a pan or rocker.

THE ROCKER.

The rocker is a box about 40 inches long by 16 inches wide and 1 foot high, with one or two riffles across the bottom, and set on rockers, as shown in Fig. 29. On the upper end is a removable hopper 18 to 20 inches square and 4 inches deep, with an iron bottom perforated with one-half-inch holes.

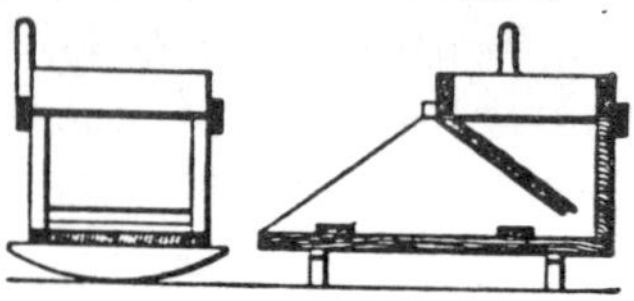

FIG. 29.

Beneath the hopper, below the perforated plate, there is a light frame placed on an incline from front to back, upon which a canvas or carpet apron is stretched. To use the rocker, material is thrown into the hopper and water is poured on with a dipper held in one hand, while with the other hand the cradle is kept rocking. The water washes the finer stuff through the bottom of the hopper, and the gold or amalgam is either caught on the apron or collects in the bottom of the rocker, while the sand or lighter material in the hopper is thrown aside. Rockers were extensively used in placer mining before the introduction of sluicing. Now they are employed in cleaning up placer claims and quartz mills and for collecting finely divided particles of amalgam and quicksilver.

THE TOM.

The "Tom" is a rough trough about 12 feet long, 15 to 20 inches wide at the upper end and 30 inches wide at the lower, and 8 inches deep. It is set on timbers or stones, with an incline of about 1 inch per foot. A sheet-iron plate or riddle, perforated with one-half-inch holes, fills the lower end of the trough, which is beveled on the lower side, as shown in Fig. 30. The material coming from the sluice, on striking the riddle, is sorted, the fine dirt and water passing through the holes, and the coarse stuff being shoveled off. Under the riddle is a shallow trough with riffles, set on an incline, into which the finer gravel passes. The discharge of the water through the plate, with the occasional aid of the shovel, keeps the sand from packing and allows the gold to settle.

The Tom succeeded the cradle rocker. The old-fashioned "Long Tom" was 14 feet long; it was followed by the "Victoria," "Jenny Lind," or "Broad Tom," 6 or 7 feet long, 12 inches wide at the upper end and 3 feet at the lower. The Tom consists of two distinct troughs or boxes placed one above the other. (See Fig. 30.) A stream of water flows in through the spout *d*, just over the place where the dirt is introduced into the upper box, or Tom proper, *a*. The dirt is thrown in by one man, while the second constantly stirs it about with a square-mouthed shovel or a fork with blunt tines, pitching out the heavy boulders and tossing back undecomposed lumps of clay against the current. To save wear and tear, the floor of the Tom is lined with $\frac{1}{8}$-inch sheet iron. The lower part of the Tom is cut off obliquely, so that the mouth may be stopped by a sheet of perforated iron, such as forms the bottom of the cradle riddle already described. The apparatus being placed on an incline, material gravitates with the water toward the sloping grating at the mouth, through which everything passes save the large stones, which gather on the

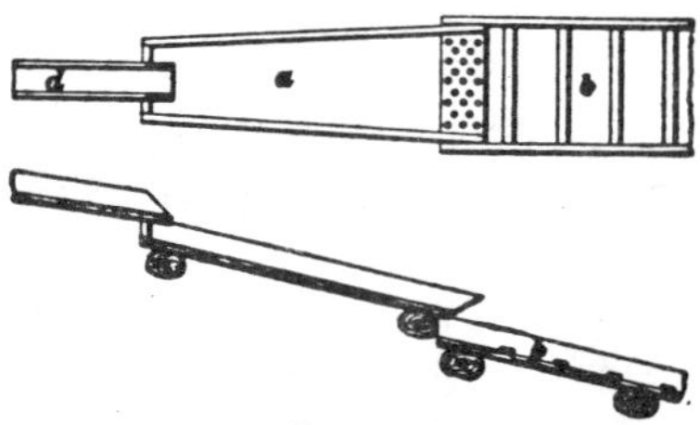

FIG. 30.

grating and are removed as often as necessary. Beneath this grating stands the riffle box *b*, into which fine material, including gold, descends. The riffle boxes, of rough plank, are also placed on an incline, just enough so that the water passing over them will allow of the bottom becoming and remaining covered with a thin coating of fine mud. In this way the gold and a few of the heavier materials find their way to the bottom and rest there by aid of the riffle bars. Sometimes a little mercury is put behind the riffles to assist in retaining the gold, and the riffle box is supplemented by a series of blankets for catching very fine gold. Toms are

FIG. 31.

cleaned up periodically, and the gold and amalgam washed out with cradles. They are applicable only to workings where the gold is coarse, as they lose considerable of the fine gold. Two to four men work at one Tom. Fig. 31 shows a Broad Tom in operation.

SLUICES.

Sluices were introduced soon after the Tom. A sluice consists of an inclined channel, through which flows a stream of water, breaking up the earth which is thrown into it, carrying away the light barren matter, and leaving the gold and heavy minerals. There are box sluices and ground sluices, the former being raised above the surface, necessitating the lifting of the pay dirt into them; the latter are sunk below the surface. Box sluices, or board sluicing, are long wooden troughs or series of troughs. They vary in length from fifty to several hundred feet, and are never less than 1 foot, and seldom more than 5 feet, wide. The usual width is 16 to 18 inches, and the height of the sides 8 inches to 2 feet. The sluice is made in sections from 12 to 14 feet long, of 1½-inch rough plank, the bottom boards sawed tapering, so that the narrow end of one box telescopes into the broad end of the next throughout the whole series, and beyond this no nailing or fixing is required. This line of trough rests on trestles, with usually a uniform grade throughout the whole series. The amount of descent, or grade, is from 8 inches to 18 inches in 12 feet. A fall of 8 inches in 12 feet is an "8-inch grade," etc. It is important that the sluice should be conveniently near the level of the ground at the point where the pay dirt is introduced; this has an influence on the grade, as has also the character of the pay dirt and the length of the sluice. The steeper the grade, the quicker the dirt is washed away by the force of the water; the tougher the dirt, the steeper must be the grade, as tough clay naturally does not break up so quickly in a slow current as in a rapid one. In short sluices the grade should be relatively light, as there is more danger of the fine gold being lost in a short sluice than in a long one. The steeper the grade, the more work the sluice can do. As ordinary pay dirt is generally completely disintegrated in the first 200 feet of a moderately low-grade sluice, the extra length is useful only for catching the gold. Sometimes, therefore, the grade of the last part of the sluice is reduced. When the grade of a sluice is very low, say 1 in 40 or 50, the gold is easily caught, and much of it would rest even upon the smooth floor of the sluice; but additional means are, nevertheless, always adopted.

When stones are plentiful in the wash dirt, a small bar may be placed across the lower end of each trough, to prevent the bottom from being run bare. It is usual to throw out stones as large as the two fists by a fork with several prongs (sluice fork). The boxes must be watched, that they do not choke up and send the contents over the sides. A false bottom is used in the sluice to catch the gold and save wear and tear on the floor proper. False bottoms are frequently made of longitudinal riffle bars 6 feet long, 3 to 7 inches wide, and 2 to 4 inches thick, two sets for each length of trough or sluice. Fig. 32 shows the arrangement of the riffle bars in the sluice. They are kept in place by cross-wedges,

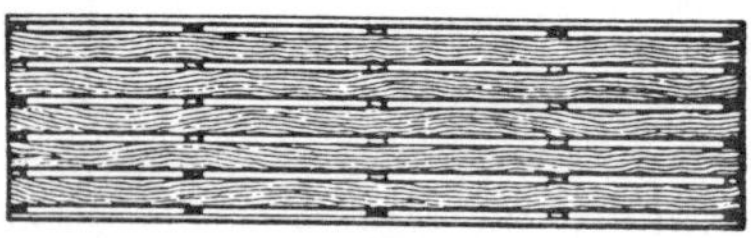

FIG. 32.

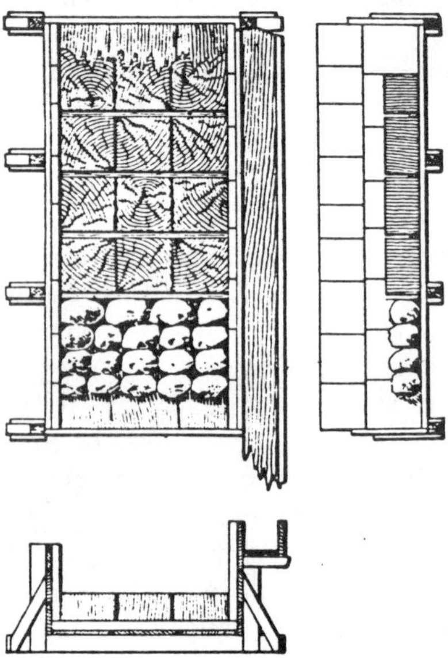

FIG. 33.

at a distance of 1 to 2 inches apart, and are not nailed, as they have to be removed at each cleaning up. Into the spaces thus formed, the gold and other heavy bodies will fall, always sinking through the lighter particles to the bottom. Where there is a great quantity of pebbles and boulders, and the pay dirt riffles described are worn away rapidly, block riffles are used, which last much longer. Instead of being sawed with the grain of the wood, they are cut across the grain, so that the fibers stand upright in the sluice-box as in the live tree. They are fixed transversely two inches apart, as shown in Fig. 33. Cobblestone and squared-stone riffles are also commonly used in such cases.

DERRICKS.

Strong derricks are used in hydraulic mining to remove heavy boulders. The style most in use at present is a mast 100 feet high set in a cast-iron box placed on sills, and hav-

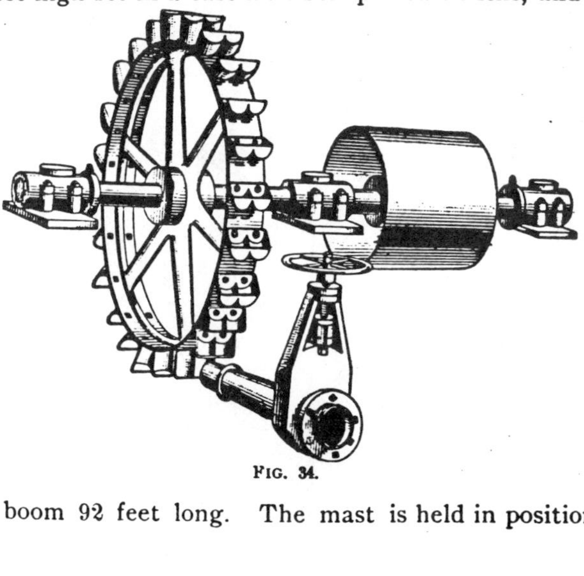

FIG. 34.

ing a boom 92 feet long. The mast is held in position by

galvanized-iron wire-rope guys one inch in diameter. A whip-block with ¾-inch steel rope is used for hoisting tackle. A 12-foot diameter hurdy-gurdy wheel furnishes the power. Stones weighing 10 and 11 tons are handled. The guys are held by double capstans, allowing the derrick to be moved without dismantling.

WATER-WHEELS.

"Hurdy-gurdy," or some other form of "impact" wheels moved by a stream or jet of water issuing under pressure from a concealed nozzle, and striking open buckets on the circumference of the wheel, furnish the power used in operating derricks and dynamos about placer mines. The Pelton wheel, Fig. 34, is the form most commonly used at present. These wheels are always housed.

CHAPTER XV.

PLACER MINING PRACTICE—BLASTING AND TUNNELING; SLUICES, UNDERCURRENTS, ETC.; TAILINGS AND DUMP.

BLASTING GRAVEL BANKS.

Where the deposits are very strongly cemented, blasting is necessary before the material can be treated. Blasting in placer mining is done on a very large scale. The usual method is as follows: The drift is run in from the bottom of the deposit a distance proportionate to the height of the bank (as a general rule not over three-fourths of the height for high banks), and the character of the ground to be moved. From the end of this main drift, the cross drift is carried each way, forming a T. The cross drifts are charged with kegs of powder, the main drift tamped by filling it up solid with dirt and rock, and the powder is exploded by a time fuse, or an electric battery. When the ground is very heavy, several cross drifts may be used. Just sufficient powder is used to thoroughly shatter the ground.

METHOD OF BLASTING.

After the loose top gravel has been washed off, the bottom cement is blasted, the object of blasting being to completely loosen the material. Hard cement requires quick powder, like Judson powder. In softer cement, black powder, as a lifter, is all that is required. In very high banks, it is best to blow out the bottom rather than to attempt to raise the superincumbent mass. The charge should then be placed so that the line of least resistance is horizontal.

In banks 50 to 150 feet high, of ordinary cement gravel,

the following method is recommended by Bowie: The main drift should be run in a distance of two-thirds the height of the bank to be blasted. The cross drifts from the end of the main drift should be run parallel with the face of the bank, and their lengths determined by the extent of the ground to be moved. A single T is usually all that is necessary. The powder required is from 10 to 20 pounds to 1,000 cubic feet of ground to be loosened.

In firing by electricity great care should be taken of wires while tamping, and where dynamite-exploders with platinum wires are used the compound circuit is desirable. In charging the drifts, the powder, in boxes or kegs, is piled in rows. Two wires *a a* and *d d* (Fig. 35) run along the middle row, the tops of the boxes on which the wires rest being removed. The exploders *b*, *b*, *b* are inserted in giant cartridges and placed on top of the paper covering the powder.

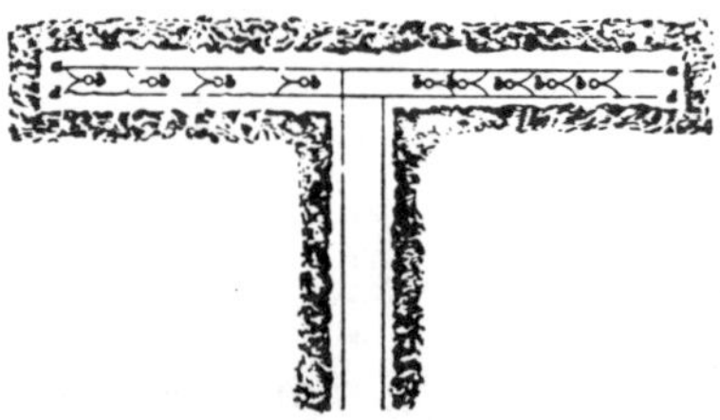

FIG. 35.

TAMPING.

Great care should be used to prevent the blowing out of the tamping. It is advisable when firing blasts by fuse to tamp nearly the entire main drift. The gravel extracted from the drifts is used for this purpose, and should be fairly dry and free from large stones. The tamping should be firmly rammed with wooden mauls, so that it will not settle. In order to prevent misfires, it is customary to lay two or three lines of fuse, which are simultaneously ignited.

TUNNELS.

Tunnels are run to open gravel claims and also to provide proper facilities for removing wash material. A tunnel should be driven well into the channel before connecting with the surface.

SHAFTS.

Shafts connecting with tunnels are usually vertical, and 3 feet by 3 feet to 4½ by 9 feet in the clear. It is sometimes convenient to have the shaft located at one side and connected with the tunnel by a short drift. Where the shaft is in hard rock no timbering is necessary, but in loose soil the shaft should be closely lagged and lined on the inside with blocks 6 to 10 inches thick, to within 10 to 30 feet of the surface. In long tunnels, a second shaft is advisable as a measure of safety.

LOCATION.

In locating the mouth of a tunnel, a point should be selected from which the sluices running in the most practicable line with a given grade can bottom the maximum extent of the pay channel at the least expense.

SLUICES.

The name "sluice" was originally applied to the miner's sluice-box. Subsequently several sluice-boxes were joined together for permanent use, and "sluice" and "flume" became synonymous. For the purpose of distinction, the name "sluice" is here applied to drifts, cuts, or boxes through which the gravel is washed, while "flume" is used solely in reference to open wooden water conduits.

Sluices should be set in straight lines, and when curves occur the outer side of the box should be slightly raised, to cause a more even distribution of the material over the riffles. Sluices with drops are effective for saving gold.

As the facility with which gravel can be moved depends upon the inclination given to the sluice, grade is very important. Where the wash is coarse and cemented and there is much pipe-clay, a heavy grade is necessary; strongly cemented gravel needs drops to break it up. A common grade is 6 inches to a box 12 feet long, approximately a 4 per cent. grade. If too much water is used, the sand packs the riffles. The best results are obtained on light grades, with shallow streams. Coarse gravel demands 4 to 7 per cent. grade and an increase of water. The water in the sluice should be 10 to 12 inches deep, covering the largest boulders. The size of the sluice depends upon the grade, character of gravel, and quantity of water. A sluice 6 feet wide, 36 inches deep, on a 4 or 5 per cent. grade, suffices for running 2,000 to 3,000 inches of water. One 4 feet wide and 30 inches deep, on a grade of 4 inches to 16 feet, suffices for 800 to 1,500 inches of water, and on a 4 per cent. grade it is large enough for 2,000 inches. As to length, the line should be sufficiently long to ensure complete disintegration of the material and allow all the gold practicable to settle in the riffles. The longer the sluice the greater the amount of gold saved; hence, the sluice is indefinitely extended so long as increased yield exceeds the additional expense.

Sluices 4 feet wide are made of 1½-inch plank, with sills and posts of 4 by 6 inch scantling. The bottom should be tight to avoid losing quicksilver, of half-seasoned lumber, free from knots, with the joints carefully sawed and grooved for the reception of a soft-pine tongue, which is inserted between the ends of the planks. Bottom and sides are spiked together, with nails 4 inches apart. Sills are placed 3 to 4 inches apart, a 4-foot sluice requiring a sill 7 feet long of 4 by 6 inch stuff. Posts are halved into the sills and firmly spiked, and every second or third post is supported by an angle brace. Bottom planks are secured to the sills by heavy spikes. Sluices should be weighted down heavily by loading the ends of the sills with stones. Fig. 33 illustrates a good construction. Each box is 6 feet wide, 12 feet long, and 32 inches deep over all. To each box are used:

8 posts, 4 in. × 6 in. × 3 ft. 2 in.
4 sills, 4 in. × 6 in. × 8 ft.
3 bottom planks, 2 in. × 24 in. × 12 ft.
4 side planks, 1½ in. × 16 in. × 12 ft.
2 top rails, 2 in. × 8 in. × 12 ft.
16 braces, 2 in. × 4 in. × 2 ft.

This is on the inside of the tunnel. On the outside of the tunnel, sills and braces are longer. The nails for the bottom are 30 penny, and for the sides 20 penny. The side lining, of worn blocks, is 3 inches thick, 18 to 20 inches deep, and is set 2½ to 3½ inches above the bottom. The riffle strips between blocks are 1¼ inches thick by 3 inches wide and 5 feet 11 inches long. The blocks are 13 inches deep and 20½ inches square. Where stone riffles are used, the bottom of the sluice is lined with rough plank. The top sluice on one side is for carrying surface water when the blocks are being set. It is 13 inches wide and 4 inches deep, made of 1½-inch plank.

RIFFLES.

Riffles are of various forms. In primitive mining, blankets, sods, hides with the hair side up, and niches cut in the bed-rock were used. Then came longitudinal strips, and finally block and rock riffles, which are now universally used. The character of the riffle used depends upon the length of the sluice, which in turn depends upon the tenacity of the material and the character of the gold, scale gold with large amounts of black sand and fine pyrites escaping all riffles for long distances.

BLOCK RIFFLES.

Block riffles are wooden blocks 8 to 13 inches deep, set on end, in rows across the sluice, each row separated by a space of 1 to 1½ inches. They are kept in position by riffle strips 1 to 1½ inches thick by 2 to 3 inches wide, held between the rows by the side lining and secured to the blocks by headless nails. Block riffles are also set and held firmly in position by soft-pine wedges driven between the blocks and the sides of the sluice, the sides or the adjacent faces of the blocks being squared. Side lining is necessary in all sluices. For handling cemented gravel the sides of the sluice should be lined with blocks 3 inches thick to a depth of 18 or 20 inches. Square block riffles are the best for saving gold, though sections of wood are frequently used just as they are sawn from the log. Rocks are the next most economical substitute, but sluices set with them require steeper grades and more water.

The life of a block depends upon the quality of the wood, the grade, the character, and quantity of the gravel, and the amount of water. The larger the amount of water on the same grade in proportion to that of gravel, the less the wear and tear on the block. Soft, long-grained wood which "brooms up" makes the best riffles. Pitch pine answers well. After each run the blocks are turned and replaced in the sluice, if not worn down too much. In repaving with old blocks, the edge worn down the most is placed up-stream. As the blocks do not fill the whole width of the sluice, the alternate rows are fitted so as to break joints.

ROCK RIFFLES.

Where heavy, strong cement is washed, stone riffles are advantageous. As quarried, the rocks are of irregular shape and size, and are set in the sluice with a slight tilt down stream. They are cheap and wear well, but have the disadvantage of being more awkward and costly to handle, requiring a longer time to clean up and repave the sluice.

A system of riffles, consisting of a row of blocks alternating with an equal section of rocks, works well, reducing the wear on the blocks, but it is not desirable for riffles which have to be frequently cleaned up. Longitudinal riffles, made of scantling placed lengthwise in the sluices, are sometimes used. Again, in some mines a portion of the line may be without a box sluice, the bed-rock being used.

BRANCH SLUICES.

Branch sluices may be necessary where a light dump requires frequent changes of the tailings discharge, or the topography of the claim is such that a single sluice can not reach all the pay dirt.

Care is necessary in turning into or turning out from a sluice, lest a gravel bar form either above or below the junction. The turn-out sluice is used when the dump room is limited; the turn-in sluice, when branch sluices are necessary to cover the whole pay area, two or more branch sluices emptying into one main sluice.

UNDERCURRENTS.

Undercurrents are introduced into a sluice line for the purpose of saving fine gold. They are broad sluices set on a heavy grade below and to one side of the main sluice. Where a drop-off can be made in the sluice, a "grizzly," made up of 1-in. × 4-in. iron bars, 10 to 20 in number, set edgewise, 1 inch apart, is let into the bottom of the sluice. It is set 1 inch below the sluice pavement, which is raised as it wears down. If too low, the grizzly is apt to clog. Coarse material passes over the grizzly, and, if the topography permits, is dropped and picked up again in sluices at a lower level. The finer stuff passes through the bars into a box about 20 inches deep, set at right angles to the main line and lined with blocks or stones. From here the material passes into the distributing box or chute of the undercurrent, as shown in Fig. 36. This chute is lined with cobbles, and provided with dividers of wood to evenly distribute the mate-

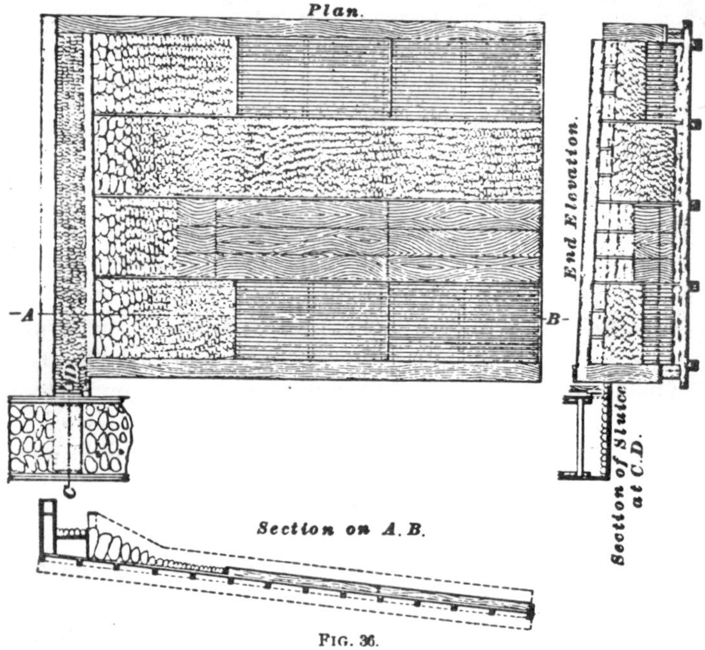

FIG. 36.

rial over the surface of the undercurrent. It has a light grade and narrows slightly towards the lower end. The undercurrent proper is a shallow wooden box 20 to 50 feet wide and 40 to 50 feet long, with sides about 16 inches high. It should be about eight to ten times the width of the main sluice. The bottom is made of 1½-inch plank, tongued and grooved and set on a grade of 8 to 10 per cent., according to the smoothness of the riffles employed. It may be paved with cobbles, wooden rails shod with strap iron, or small wooden blocks. With the smooth riffles a grade of 12 inches in 12 feet is plenty, but with blocks the grade should be increased to 14 inches in 12 feet, and with cobbles to 16 inches. The tailings from the undercurrent are led back into the main sluice. At French Corrall, with a tailing sluice 5 feet wide, the yield of the first undercurrent, which was 20 feet in width, was 20 per cent. of the total yield of the undercurrents; an addition of 10 feet to the width increased its yield 27 per cent. of the total, without changing the grizzly in the main sluice.

TAILINGS.

The refuse from any form of mining, after the extraction of the gold, is called "tailings." Placer refuse is also known as *débris*. Tailings from mills consist of pulverized quartz or other gangue matter. Refuse from gravel workings is of all sorts and sizes of material. The light particles of soil, loam, and sand are easily carried forwards by running water, while the rocks and boulders, though readily transported through the sluices, lodge and distribute themselves when discharged therefrom into the creeks and streams, in accordance with their size, shape, and specific gravity, and for their removal time and flood are necessary.

THE DUMP.

One of the requisites for a large placer enterprise is a good dumping ground. As thousands of tons of material are being treated daily, and only a very small portion of this is saved, some place must be provided at a lower level for the storage of the waste material. A much larger space is necessary for this purpose than was originally occupied by the excavated material. The lack of dump room is remedied only in exceptional cases by discharging the waste into a current or torrent. This may occur where the gold placers are on the borders of large, rapid, well-confined streams, but in the mountains, where the majority of gold deposits occur, the rivers are narrow and shallow.

CHAPTER XVI.

PLACER MINING PRACTICE—WASHING OR HYDRAULICKING.

CHARGING THE SLUICES.

The tunnel and sluices being finished and everything in readiness, water is turned into the pipes. The sluices are run for a day in order to pack them; the water is then shut off and a charge of quicksilver is put into the upper 200 or 300 feet of the sluices and a small quantity distributed along the entire line with the exception of the last 400 feet. In a 6-foot sluice the first charge should be about 3 flasks. The undercurrents are charged at the same time, and a little quicksilver put into the tail sluice. Quicksilver is added daily during the run, in gradually lessening quantities, the object being to keep the mercury uncovered and clean at the top of the riffles; therefore, the charge is regulated by the amount exposed to view. A 24-foot undercurrent requires 80 to 88 pounds of quicksilver.

In charging the riffles, the quicksilver should not be sprinkled or splashed, as particles of mercury are readily carried off by the swift stream, while the very fine portions will even float in still water. Top water from mining sluices often yields minute particles of quicksilver.

CAVING BANKS.

The first work is started near the head of the sluice. As the bank recedes, the bed-rock cuts are carried along with it and the sluice advanced. In order to cave a bank, two streams of water are usually delivered against its base at an angle with each other, as in Fig. 37. This great mass of water, discharged against the bank with a force of 150 to 200 pounds per square inch, rapidly undermines it, the bank crumbles away, and the rush of water carries the débris into

FIG. 37.

the sluices. If the dirt caves rapidly, one pipe may be used for cutting, while with the stream from the other, the falling gravel is washed into the ground sluices. The face of the bank should be kept square; advantage should be taken of such corners as are left, and under all circumstances a "horseshoe" form should be avoided. When the cut is rapidly pushed ahead and the work not squared, the men at the pipes are encircled by high walls, and their lives are in danger. Where banks exceed 150 feet in height, the deposit should be worked in two benches. When the men at the pipe see the bank is about to cave, the water should be immediately turned away, for if the cave falls on the water in the ground cut, a rush of débris follows, and the men at the pipe have to run for their lives. Caves are usually made towards evening. Locomotive reflectors or bonfires illuminate the banks at night; electric lights are also largely used around the larger placer mines. The washing should be continuous, and no water allowed to run to waste; therefore, it is desirable to have several faces or openings, so that the stream may be diverted from one to the other while cuts are being advanced and sluices lengthened. These cuts, or *ground sluices*, are trenches made in bed-rock near the face of the bank to collect the water and material and convey them to the sluices. Sometimes they are 60 to 70 feet deep.

As a precaution against theft, where claims are worked intermittently, sluices are run full of gravel before closing down.

CLEANING UP.

The length of a run depends largely upon the wear of the pavement. Some claims are cleaned up every 20 days, others every two or three months, and a few only once a season. All pavements should be cleaned up as soon as they begin to wear in grooves. Where a large quantity of gravel is washed, it is advisable to clean up the first 1,000 or 2,000 feet of the sluice about every two weeks. The tail sluices are cleaned up only once a year. Undercurrents should be cleaned up whenever quicksilver is found spread over their lower riffles, with a tendency to discharge over their ends.

When it is decided to clean up, the bed-rock and ground sluices are washed clean. No material is turned into the sluices, clear water alone being run in until the sluices are free from dirt. A small quantity of water, in which a man can conveniently work, is then turned through the sluice, and the blocks are taken out by crowbars, washed clean of amalgam, and laid alongside the sluice. This is done in sections of 100 feet. One row of blocks is left in the sluice between each section; these rows serve as riffles to prevent the gold and quicksilver passing down the sluice. After the first section of blocks is taken up, the men follow the gravel and dirt as it is slowly washed down the sluices and pick up the quicksilver and amalgam with iron scoops and deposit it in sheet-iron buckets. As each riffle is reached, the amalgam and quicksilver are collected, the block riffles removed and the residue washed down to the next riffle, and so on down the entire line of sluice. When this operation is finished, the water is turned off entirely, and workmen go over the sluices with small silver spoons, digging the amalgam out of nail holes and cracks. After this, the side lagging is overhauled, and the blocks are then replaced. Very long sluices are usually lined in the lower portion with heavy rock riffles, which can be used for longer periods without cleaning up. It is customary, where mines are run night and day, to clean up as long a section as possible during the day, and to replace the lining and resume washing at night, proceeding thus till the whole is cleaned up.

AMALGAMATION.

Though heavy gold may be arrested by the various contrivances described, much fine gold would escape in the absence of mercury or quicksilver. If this is present, however, it instantly seizes and amalgamates any gold coming in contact with it. When using zigzag riffles, a vessel containing quicksilver and pierced by a small hole which allows the metal to escape drop by drop is placed near the head of the sluice. Trickling down from riffle to riffle, it overtakes the gold, absorbs, and retains it, the amalgam thus formed being caught in the longitudinal or block riffles farther down. In longitudinal riffle sluices, after starting the washing, some of the mercury poured in at the head of the sluice finds its way down with the current, but the larger portion remains in the upper boxes. Smaller quantities are introduced at intervals lower down, the quantity being increased in direct proportion to the amount of fine gold present. Another plan is to impregnate with mercury the pores of the wood forming the riffle bars by driving a piece of gas pipe, ground thin at one end, into the wood and filling it up with quicksilver. The pressure of the column forces a certain amount into the fibers of the wood. This catches the gold, and the resulting amalgam needs only to be scraped off the surface of the wood.

A fourth device, for use where there is very much fine gold, is the amalgamated copper plate. This usually measures 3 feet wide by 6 feet long. Sometimes the stream is split and carried over two or three separate plates. The plate is placed nearly level and at a considerable distance from the head of the sluice, as it is intended to catch only the fine float gold, and for this reason, also, a sheet-iron screen, perforated with holes $\frac{1}{2}$ inch by $\frac{1}{16}$ inch, is placed in front of it, so that only the finest particles pass over it. It is amalgamated by first cleaning its upper surface with weak nitric acid and then applying some mercury, which has been treated with dilute nitric acid to form a little nitrate of mercury. The current must be slow and shallow, so that every particle of gold may come in contact with the face of the plate. A freshly amalgamated plate may become coated with a green slime of subsalts of copper; this must be carefully scraped off and the plate rubbed with fresh mercury. To remove the amalgam, the plate is taken up and gently heated, and it may then be easily scraped off. The plate is allowed to cool and again rubbed with a little mercury. The plate should not be less than $\frac{1}{16}$ inch thick.

AMALGAM KETTLES.

Amalgam kettles are ordinary sheet-iron buckets or porcelain kettles. In cleaning up they are used as receptacles for floating the gold amalgam, which is floated in quicksilver to free it from barren substances before straining and retorting.

CLEANING THE AMALGAM.

The quicksilver and amalgam obtained in cleaning up are well stirred in buckets, and the coarse sand, nails, and other foreign substances which float to the surface are skimmed off. This residue, which retains considerable amalgam, is concentrated by working in pans or rockers, and the concentrates are ground in iron mortars with some clean quicksilver. Any base material floating to the surface of the bath is melted separately to a base bullion; the remainder is added to the fine amalgam. The quicksilver is strained from the amalgam through canvas or drilling, and the dry amalgam is treated in iron retorts.

RETORTING.

When the amount of amalgam to be treated is small, the hand retort answers all requirements; but at large gravel mines, stationary cast-iron retorts are used. When large quantities of amalgam are retorted and the furnace is left unattended, a retort which is set immediately above the fire is apt to become overheated. The weight of the metal inside of the retort then causes it to "belly," ruining it completely. To prevent this, the retort should be supported at several points and arranged with the fire to one side, so that the heat may be evenly distributed over it. (See Fig. 38.)

Before putting the amalgam into the retort, the latter should be coated on the inside with a thin sheet of clay, which prevents the amalgam from adhering to the iron. The amalgam should then be carefully introduced and spread evenly. The pipe connecting the back of the retort with the condenser must be cleared of all obstructions, and the amalgam should be so spread that by no possible mischance can this pipe become choked, as an explosion would probably result, filling the retorting room with the poisonous

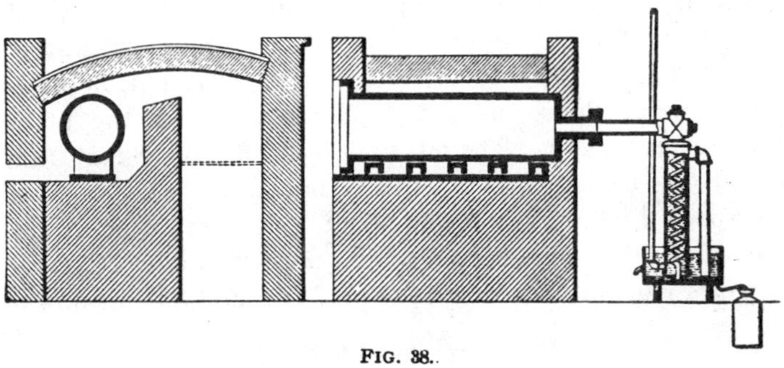

FIG. 38.

fumes of mercury and greatly endangering its occupants To avoid danger, the heating should be very slow at first. After the cover has been put on with a luting of clay or a mixture of clay and wood ashes and securely clamped, the fire is lighted and the heat gradually raised, a dark red heat being all that is necessary to volatilize the quicksilver. Towards the end of the operation, the heat is raised to a cherry red, until distillation ceases. The retort is then allowed to cool, and, when cold, is opened. During the operation, the condensing coil at the back of the retort should be kept cool by a continuous supply of fresh water entering from the lower end of the box which contains it, while the discharge of warm water is effected above. The retort bullion is cut or broken into pieces and melted in a well-annealed black lead crucible, and the gold cast into bars.

DISTRIBUTION OF GOLD IN SLUICES.

In sluicing, the greater part (usually about 80 per cent.) of the gold caught is found in the first 200 feet. For example, of a claim yielding $63,000 on a hundred days' run, $54,000 was obtained in the first 150 feet, and $3,000 from the undercurrents. The first undercurrent, 790 feet from the head of the sluice, yielded 50 per cent. of the total amount taken from the undercurrents. The second undercurrent, 78 feet distant below the first, with a drop of 40 feet between them, contained 33 per cent. of the gross undercurrent yield. The last undercurrent was 98 feet from the second, with a drop of 50 feet between them; its yield was about $500. Sometimes about a hundred feet at the head of the sluice is covered with gravel during the greater part of a run; in such cases, the gold is found farther down.

LOSS OF MERCURY.

A certain loss of quicksilver is unavoidable in placer mining, the amount depending on the grade, length, and condition of the sluices, the character of the material washed, the amount of water used, and the length of the run. The loss may be reduced by lengthening the sluice line, keeping joints tight, and careful cleaning and chinking. On large enterprises it is usually in the vicinity of one pound of mercury for every $100 worth of gold recovered.

LOSS OF GOLD.

The loss of gold is inversely proportionate to the size of the grains, the length and grade of sluices and undercurrents, the depth of the water, and the completeness of the breaking up of the pay dirt. Frequent drops in a line assist in disintegrating the gravel and allow of shorter sluices. In washing hard cement banks, it is advisable to use plenty of powder to thoroughly shatter the bank, and large lumps of cement should be broken up before being introduced into the sluice. However carefully the operation be conducted, there is invariably some loss of fine gold; the last under-current will always catch some gold, and the tailings will show a trace.

CHAPTER XVII.

EXAMPLES OF PLACERS—THE ALMA PLACER.

As an example of an ordinary hydraulic placer mine, we may take that of the Green Mountain Company, at Alma, South Park, Colorado. In South Park, at an altitude of 10,000 feet above the sea-level, is an extensive area of placer ground, located along the banks of the South Platte River, and extending from the base of Mount Lincoln to Fairplay, a distance of over 20 miles. This area consists of rolling banks of pebbles, boulders, gravel, and sand on both sides of the stream, covered with grass and a few spare trees, and sloping up gently towards the mountain sides for an average width of about half a mile. Portions of these placer banks have long been worked, both at Alma and Fairplay, but the banks are far from exhausted. The principal hydraulic workings are at Alma, where also the banks are thickest, owing to the confluence of tributary canyons and streams at that point. A powerful body of water is at hand during the summer months, and the beds are worked continuously, night and day, during the season.

The main source whence the gold originated was doubtless in numerous large, partially developed quartz veins in granite, at the head of the ravine above Montgomery, at the foot of Mount Lincoln, where are the headwaters and main sources of the South Platte River. Besides these gold-bearing veins, the quartzites and porphyries of the adjacent region may have contributed a certain amount of gold to the placer from gold disseminated generally throughout their mass. The head of the canyon below Mount

Lincoln was the starting point of the glacier that carved out the valley upon which the Alma placers lie, the line of which is now occupied by the Platte River.

The character of the predominant pebbles in the placers —quartzites, granite, and porphyry—suggests the rocks at the head of the canyon as the principal source of the gold. The summit of Mount Lincoln is 14,400 feet above the sea-level, and about 4,000 feet above the valley of the Platte. The east face of the peak descends in a steep cliff of massive granite, capped by quartzites and limestones carrying inter-bedded sheets of porphyry. The face of the granite cliff is traversed by a great number of wide, parallel fissure-veins of quartz and feldspar, carrying more or less gold and pyrites. The valley below is U shaped, betokening the pathway of an ancient glacier, and scooped out of granite by the ice. The rocks over which the glacier passed in its downwards course are rounded, polished, and grooved, forming what are called "glacial sheepbacks." These form the pavement of the upper part of the ravine. Near the head, a violent stream, whose source is in a small lake in a glacial amphitheater still higher up, descends in bounding waterfalls—the source of the Platte. Below where the falls plunge into the valley is a small, shallow lake, half filled with gravel, supposed to contain much gold scooped out from the veins in the upper part of the canyon, and a scheme is projected to drain the lake and work the gravel by a coffer-dam and underground sluice tunnel, the water and débris from the washing passing through the coffer-dam and out through the sluices in the tunnel. The hollows at the base of the waterfall, contrary to what might be supposed, are not found to be very productive. From the lake, we look down on numerous traces of the work and pathway of the ancient glacier. Vast bodies of huge boulders rise on the slopes at the base of the mountain to a height of nearly 1,000 feet above the stream, with here and there an exceptionally large block dropped by the melting ice on the top of the moraines on either side of the stream. Between the moraines, the river runs through a long meadow, with continuous banks of placer material on either side from 50 to 200 feet in height. The surface of these morainal banks is undulated, rising and falling in smooth, grassy swells, like the waves of mid-ocean. These banks are composed of "modified drift"; that is, the rough, angular blocks left by the glacier have been worked over by the stream, pounded and broken up, and in part reduced to gravel. When exposed in section, as at Alma, they exhibit the structure from surface grass roots down to bed-rock shown in Fig. 39.

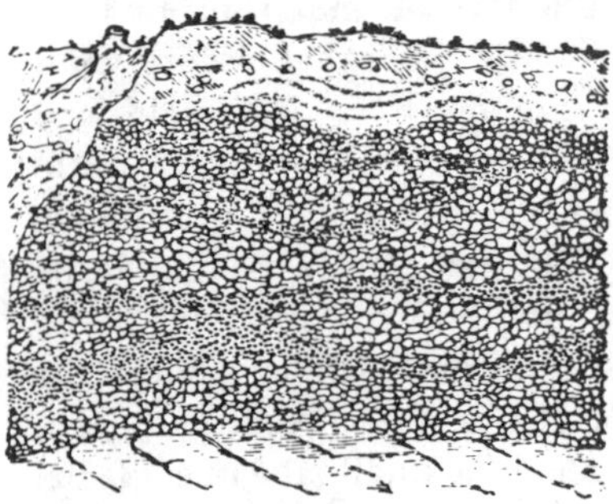

FIG. 39.

The structure in detail is as follows: First, a foot or two of black turf, in which there is little gold; below that, a foot or two of clay with pebbles in it, and then a few feet of sandy layers, irregularly bedded, in dovetailing streaks, as if formed by eddies and currents, and, likewise, comparatively poor; the remainder, to bed-rock, 30 to 50 feet, is composed of subangular and rounded pebbles and boulders of all sizes, from a fraction of an inch to a yard in diameter, cemented together by gravel, sand, clay, and, in places, by iron oxide, into a tolerably fine conglomerate, which can only be successfully attacked by the point of the pick or the all-destroying Giant nozzle. These banks are continuous down both sides of the creek for several miles, and are thickest at Alma, opposite the outlet of the tributary canyons, Buckskin and Mosquito. Here is the site of one of the oldest working placers in Colorado. The banks have been cut back for a long distance from the river, presenting a face of vertical cliff 70 feet in height and about a mile in length, channeled by narrow ravines and gashes, from the inroads of the Giants and the cutting back of ditch and flume waterfalls. Some of these cuts are short, narrow gashes, not penetrating far into the hills; others lead, through long, narrow, ravines into wide, open amphitheaters surrounded by channeled cliffs, while the center is occupied by piles of large boulders thrown out from the sluices and stacked up in the course of the work. Winding through these paths of debris may be seen the remains of the old, abandoned gravel sluices, telling of work done long ago.

To enter one of these amphitheaters, where the work is still actively progressing, we approach by way of one of the ravines penetrating the hill. From this ravine issue two long,

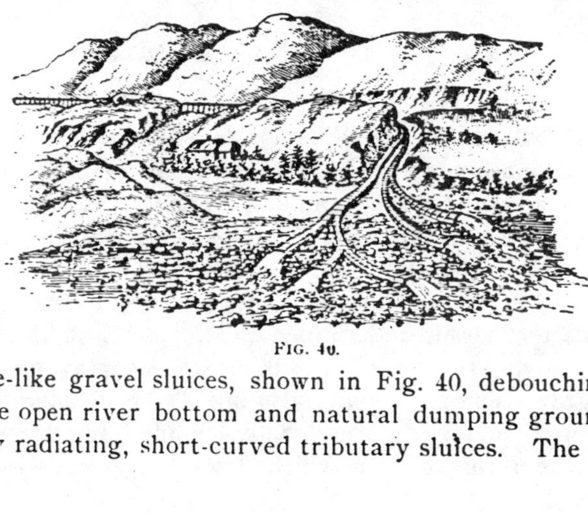

FIG. 40.

snake-like gravel sluices, shown in Fig. 40, debouching on to the open river bottom and natural dumping ground by many radiating, short-curved tributary sluices. The water rushes with great speed and force along the wooden bottom of the sluices, and the bigger boulders can be heard rolling and bumping over the wooden riffle blocks which pave the bottom. We follow up these sluices through the ravine for over a thousand feet, till it opens in a broad amphitheater 200 feet wide by 70 feet deep. Here operations are in full blast. Several flume waterfalls, shown in Fig. 41, descend the steep bank at the head of the amphitheater, at varying distances apart, each one cutting back rapidly a sharp, narrow ravine or channel for itself from grass roots nearly to

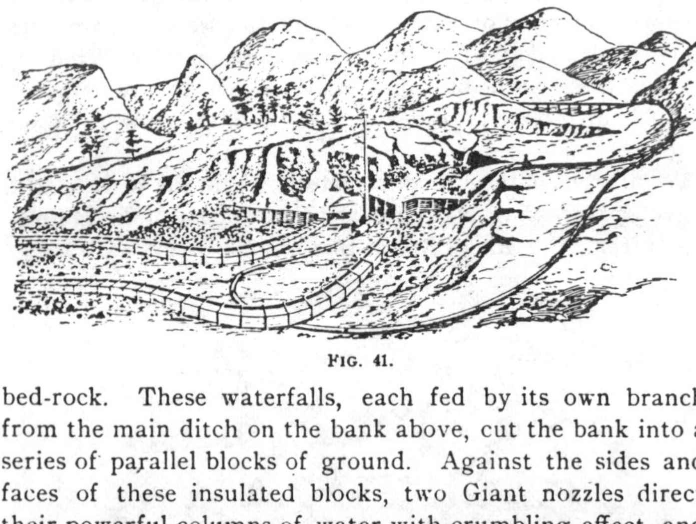

FIG. 41.

bed-rock. These waterfalls, each fed by its own branch from the main ditch on the bank above, cut the bank into a series of parallel blocks of ground. Against the sides and faces of these insulated blocks, two Giant nozzles direct their powerful columns of water with crumbling effect, and the partially cemented material fades rapidly before them as mass after mass is undermined and topples into the refuse stream, and thence is hurried into the gaping mouths of the gravel sluices. Giants also speed the boulders and gravel on their way by occasionally lending their force to that of the refuse stream which flows from beneath the waterfalls. Thus sand, gravel, and boulders are washed into the gravel sluices, the bottoms of which are lined with riffles of short cross-sections of the trunks of pitch-pine trees, placed close together, like rows of lozenges, or like a Nicholson block pavement. Both big and little boulders and gravel roll rapidly over this block pavement, and the gold, by its gravity, drops to the bottom and is caught between the interstices of the blocks and retained there. Its retention and deposition are further aided by throwing in quicksilver, which, by its affinity for gold, collects the finer particles in its soft, heavy, silvery mass. While the boulders and gravel soon find their way to the natural dumping ground on the open river bottom, the gold in its travel stops long before that point is reached.

In the center of the amphitheater, a tall derrick, driven by a 10-foot Pelton wheel with an undershot nozzle, moves a long arm slowly around over the area. The use of this derrick will presently be apparent. One of the flumes having been stopped and its attendant waterfall having ceased, and the Giant nozzle being directed elsewhere, the pathway of the refuse stream becomes comparatively dry. Then men climb into it and pick out the larger boulders, too large to pass through the gravel sluices—some of them being so large as to require blasting. Then the long arm of the derrick swings around, and the boulders are piled into a large stone-boat and carried around to a convenient dumping ground on either side of the gravel sluices. The largest boulders being thus removed, the gravel and small pebbles become more exposed, and the Giants are again brought to bear on these, till at last bed-rock sandstone appears, full of cracks and crevices, forming by its gentle dip and inequalities natural riffles and lodging places for a portion of the gold. The bed-rock cleaners now dig up and shovel into the sluice the rotten surface of the sandstone to a depth of a foot or so, or to such a depth as experience has proved that gold occurs. They probe the cracks in bed-rock with their knives and brush the rocks with small brushes and pick out any stray nuggets that may be concealed. In the bed of the stream descending from the flume, men are also at work with long-handled shovels, ground sluicing, or helping along and removing out of the way some of the boulders, so as to keep the water in as definite a channel as possible and prevent it from spreading. The ravine, which is 1,000 feet long, and the amphitheater, 200 feet wide by 70 feet deep, were both excavated within six months.

PRELIMINARY WORK AND PROSPECTING OF THE PLACER.

Before undertaking this enterprise, the ground was well prospected and the presence of gold in paying quantities assured. Shafts and prospecting holes were dug down to bed-rock to ascertain the depth of the formation. Prospecting by panning was also carried on along the exposed sides of the gulches. The water supply was considered, and the ditch and flume planned with a view to its power over the underlying bed-rock. The grade of the ditch was also considered, for if the grade is too great the water cuts and breaks its banks. Three-eighths of an inch to a rod was found to be a good grade. Penstocks and boxes were made, and pipes 14 inches in diameter attached. The Giant nozzles, having been attached to the pipes and firmly braced to strong wooden platforms on the ground, began to play on some natural exposure of the bank, while the ditch flumes commenced their work of cutting back ravines and blocking off the ground to be later broken down by the Giants. The gravel sluices were constructed for carrying the pebbles, gravel, and gold, with a general grade or inclination towards the natural dumping ground on the river bed, and the rest is as already described.

RELATIVE RICHNESS OF DIFFERENT ZONES OF THE BANK.

The richness of the bank appears to depend upon various conditions. The fine, eddy, top sand is seldom rich, the best gold being in the coarser material or on and in bed-rock. The gravel is sometimes cemented by iron rust to the consistency of rock. Black sand occurs here and there, richest where rusty. There are often peculiar courses in the sand currents, and turnings and windings as in river courses. As many as three different periods of deposition of gravel may be observed.

RESERVOIR, DITCH, FLUMES, ETC.

The reservoir up the river, supplying the ditch, covers about 5 acres and is 10 feet deep. The dam is made of gravel and brush, cribbed with timber and having a gate. The ditch that leads to the highest gravel banks is two miles long and carries about 2,000 miner's inches of water. It is 12 feet wide and 3 feet deep, and flumed on trestles, at one place, for 240 feet. The flume is of boards 12 feet in length, of sawed pine timber, forming boxes, built with frames 4 in. × 4 in.; the floor boards are 1½-inch material, sides 1¼ inches; the flume is 6 feet wide and 3 feet deep.

At the end of the wooden flume, on solid rock, is a flume 50 feet long, at right angles to the main flume. From this there are four openings to smaller ditch flumes, which distribute the water to the general workings. The grade of the ditch is three-eighths of an inch to the rod. From the main ditch, a branch ditch leads to the penstock or sand box. From that, two pipes are laid, which, at the penstock, are 22 inches in diameter, but taper gradually toward the Giants to 10 inches in diameter. These pipes are each 500 feet long. There are two Giants, of the size known as No. 2. The discharge pipe of these is 9 feet long. The deflector, by which the man in charge directs the nozzle in any direction he pleases, is screwed on the end. The deflector works on the principle of a ball and socket; where the discharge pipe connects with the main casting there is also a ball and socket, so that it, too, can be moved to right or left, and up or down. Leather is used to prevent leakage at the joints, and sawdust is thrown into the sand box to stop leaks in the pipes. Giants, Chiefs, or Monitors, as they are variously called, are used for cutting down the banks. The water that is not used by the pipes is allowed to run over the highest part of the gravel bank to cut down and carry away gravel to the sluices. The pipes use 200 inches of water to each Giant. The ditches carry 2,000. The volume carries the gravel into the sluices. The sluices are 3 feet wide by 4 feet high, or deep, paved with round block riffles 8 inches thick and of varying widths, packed on the bottom of the sluices with small pieces of rock. (Fig. 42.) So great is the force of the water in these sluices, that boulders 100 pounds in weight are sometimes carried from end to end. The velocity is about 25 miles per hour. The dip, slope, or grade is 4 inches to every 12 feet, or 33⅓ inches to every 100 feet. The sluice is laid on bed-rock, which is sometimes cut down to admit it. The curves of the gravel sluices are made like those of a railroad, raising the outer side of the curve. There are two parallel sluices 30 feet apart. When these two main arteries

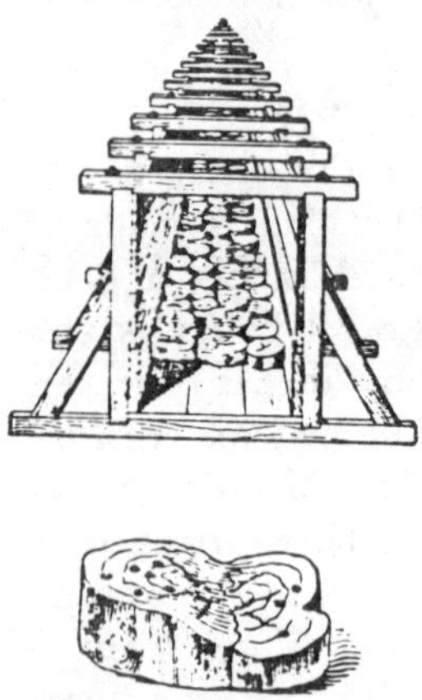

FIG. 42.

reach the bed of the river, which is their natural dumping ground, branches are formed so as to spread out the material in a fan shape, and these branches are extended as the material accumulates. (Fig. 40.)

The sluices are 4,000 feet each in length. The riffles protect the boards in the bottom from wear and tear of the gravel and boulders. Old riffles are left in the bottom of the branches where no gold is collected, for this purpose. The gold is mostly found deposited in the first 400 feet of each sluice. The derrick or hoisting gear is run by water, hoisting big rocks in a stone-boat by a gin block and chains. Water is led by an 8-inch pipe from the sand boxes to an undershot Pelton wheel with 1¾-inch nozzle. The wheel is 10 feet in diameter, and the drum works on the **V** principle.

CLEANING UP.

About 2 ounces of quicksilver to each ounce of gold in the riffles is thrown into the flume. In a clean-up, which occurs at regular intervals, the riffles are first taken out and then water is let on to wash everything clear. The packing of small rocks is taken out with twelve-tined forks. The floor is then cleaned, the gold usually being all collected by the quicksilver at about 200 feet from the entrance of the flume. The quicksilver is shoveled out, separated from black sand, and carried in kettles to the retorting office; there it is retorted and prepared for the mint.

CHAPTER XVIII.

EXAMPLES OF PLACERS—THE ROSCOE PLACER.

As an example of placer mining by turning the course of a river by means of wide flumes, and leaving the river bed dry and bare for a space, for operations down to bed-rock, we may cite that of the Roscoe placer in Clear Creek Canyon, Colorado. Clear Creek Canyon is one of the steepest and grandest of the Frost Range. It is cut through granite rocks for a linear distance of some 40 miles, to an average depth of 1,000 feet. About 13 miles from its outlet on the plain, the creek forks, one branch leading up towards the gold-mining region of Central, the other to the gold and silver region of Idaho Springs. The main creek receives the drainage of two gold-bearing districts. At Central, in addition to gold from the mines, veins, and rocks direct, the creek brings down a great deal of flour gold, the refuse of old stamp mills, which by their crude methods lost in the past upwards of 40 per cent. of gold, together with a great deal of amalgam. This refuse matter has been accumulating from the mills alone for the past thirty years, not to mention what has been derived from the rocks themselves by the ordinary process of nature. Miners and prospectors in past times obtained a great deal of gold from shallow surface washings, without attempting to reach the deep-lying but coveted bed-rock, where the most of the gold was reasonably expected to lie.

Near the location of the placer, the canyon is at its deepest and narrowest. Several huge veins of quartz and feldspar, doubtless carrying more or less gold, cross the canyon from side to side. It is by the erosion and breaking down of these great veins, which originally stretched across the canyon like a dyke, that we enter the grand portal leading to the Roscoe placer. Originally, the great feldspar vein stood as a natural dam across the waters of the creek, until they undermined and broke through it, and the vein collapsed into the creek in huge boulders, over which the waters now dash in foaming waterfalls, with a sudden drop of 30 feet.

This natural dam was selected as an excellent point for dumping the material to be dug out immediately above it. The débris thrown over this fall would be rapidly disintegrated and carried down stream by the torrents; so nature supplied one of the first requisites for enterprises of this sort—a good dumping ground. Above the stone dam is a stretch of a couple of miles of comparatively slow-moving and shallow water in a natural widening of the creek bottom, underlaid by deep gravel. The railroad runs on the bank, convenient to the placer, and the grade was also convenient for laying alongside of it the pipe lines to run the Giants, etc. The opposite bank was low, and the slope gentle and well adapted for constructing the great flume and ditch to carry off the waters of the creek.

PRELIMINARY PLAN AND WORK.

Before commencing operations, the ground was prospected by shafts down to bed-rock, and the presence of gold assured. The general plan of the work to be done was as follows:

First, a wide and long flume was to be constructed on the south bank, capable of carrying off all the water of the creek, which was to be turned into it by means of a dam, laying bare a mile or more of the river bottom. At the lower end of the property, just above the stone dam, a pit was to be dug to bed-rock. In this a Ludlum gravel lifter was to be placed—a large funnel-shaped pipe, up which water, gold, gravel, and stones would be carried by the force of a Giant nozzle below it into an elevated gravel flume on the surface above—to which would also be attached several undercurrent sluices for catching the finest gold. A pipe line, some two or three miles long, with a head of over 100 feet and carrying 1,000 inches of water, would give the needed power to the Giants in the pit. As the workings would advance up the creek, the abandoned pit would receive the dump of the work in progress.

BUILDING DITCH AND FLUME.

Two things had to be done simultaneously: one, to build the big ditch and flume to carry off the water of the river and leave the bed dry; the other to build a pipe line to get sufficient head of water to work the nozzles and sand pumps at the places chosen for excavation. By a natural widening of the river bank and its encroachment on the stream at one place, the channel of the stream is locally contracted. Starting with this natural advantage, a temporary dam and flume was built of sacks filled with sand, to keep back the water till a more substantial "triangular" dam, of timber partitions filled with stones, could be built. Thus a ground flume was constructed, as shown in Fig. 43. First, a pile or wall of sand-bags next to the water of the ditch, and then behind that a framework of timber with triangular partitions filled with

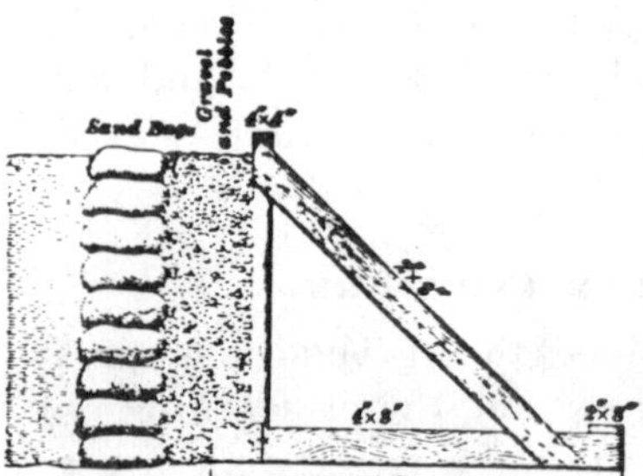

FIG. 43.

FIG. 44.

stones and pebbles, faced, or "rip-rapped," on the outer side with heavier stones, until the nature of the ground should require a flume of sawed timber to be constructed.

This flume that carries off the river water is 10 feet wide, 6½ feet high, and 2,600 feet long, with a capacity of about 32,000 gallons per minute. The bents are 4 in. × 8 in. and 16 feet long, with braces on the outer side at an angle of 11½°. The braces are 2-in. × 8-in. plank, 5 feet long, bolted to the 4-in. × 8-in. sill and upright post. The flooring is 4 inches thick, the boards 12 inches wide and 16 feet long. The flume is not straight, but follows the course of thestream, the floor being elevated on the outside of the curves an amount corresponding to the degree of curvature, as on a railroad track, making the water run level. The grade on the curve is 1¼ inches to 16 feet. When the flume is straight, the grade is ¾ inch to 16 feet. The angle at which the floor is cut for joining is not over 30 degrees. The sides are made of 2-inch boards 16 inches wide.

WATER-POWER AND PIPE LINE.

The next matter to be attended to was to get sufficient head of water for the Giants. To effect this, they had to go three miles up the river to a point where the descent of the stream was somewhat steep and rapid. There they built an intake flume of wood, 6 feet wide, 4 feet deep, and 800 feet long, to a penstock or sand box connecting with a wooden-stave pipe 48 inches in diameter at its widest point. To enter the penstock the water passes through a screen or iron grating, which catches the coarse rubbish, such as leaves, sticks, etc., floating in the water, and the overflow passes through gates on the south side. The main current passes into the penstock, which is 8 feet square and 16 feet deep. At the bottom is a well which collects any débris, so the

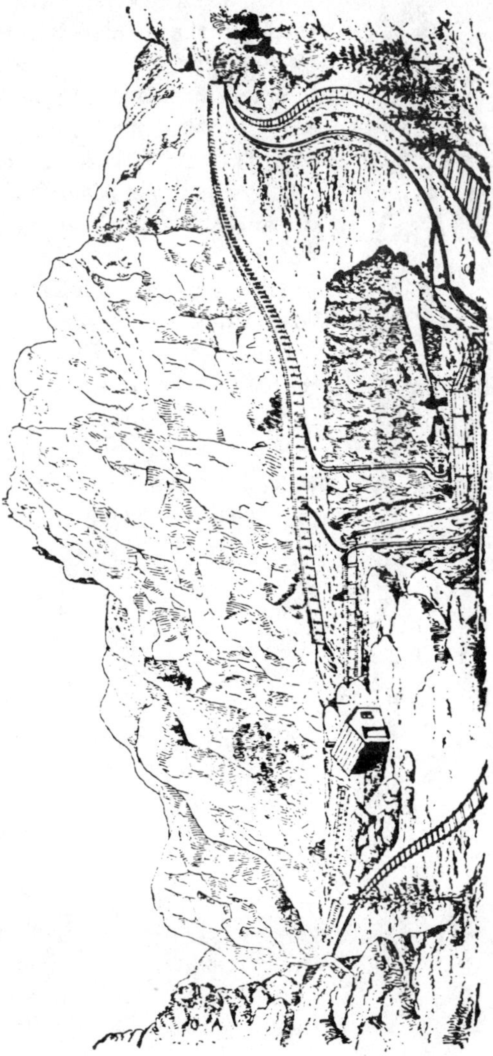

FIG. 45.

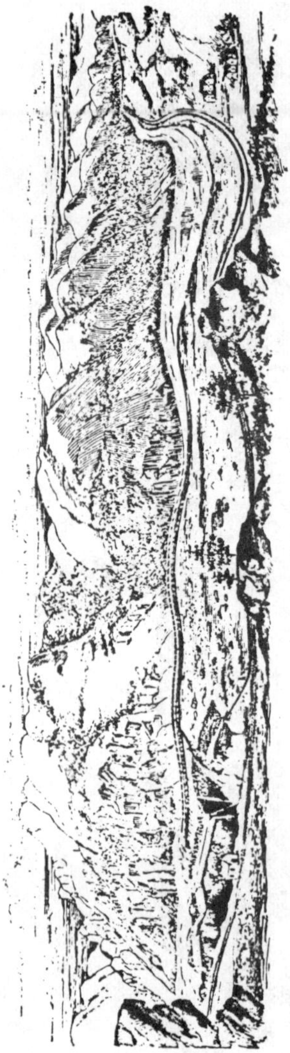

FIG. 46.

water passes clean and pure through the penstock into the 48-inch pipe. This pipe is made of staves or boards of pine, banded with round steel hoops. (See Fig. 44.) After leaving the penstock, the pipe is buried for a distance of about 100 yards under a stone embankment, and passes by an arch under the railroad track to its junction with the metal pipe. Where it has to withstand the greatest pressure, it is closely banded. The pipe diminishes gradually in diameter, till, from 48 inches at the penstock, it becomes 22 inches at its junction with the metal pipe, which also, in its course, diminishes to 16 inches. The steel pipe is three-eighths of a mile long. A still smaller pipe, 12 inches in diameter, connects with this and runs parallel with the main pipe, forming two pipes, for one-eighth of a mile. One of these pipes is for the Giant nozzle, the other to supply the sand pump for elevating the gravel from the bottom of the excavation into the gold gravel sluices. The pressure on these pipes at the nozzle is 87 pounds per square inch, and they will throw a column of water 165 feet from a nozzle 4 inches in diameter. With a closed pipe the pressure would be 189 pounds.

Fig. 45 gives a general idea of the lower end of the works, and Fig. 46 shows a panoramic view of the whole plant, from the intake flume, far up the canyon, to the penstock, and from the penstock along the line of big pipes to their final connection, at the lower end of the placer, with the Giant nozzles, and on the opposite side of the river the big flume carrying the water of the river out of its natural course and leaving the bed dry for operations.

The pit, above the stone dam, is also shown, as completed, in Fig. 47. Giant nozzles play against the sides and into the bottom of this pit, washing down the débris of the banks

FIG. 47.

and excavating the bottom, while gravel elevators and water-lifters force up the material excavated to an elevated sluice, to be winnowed of its coarse gold; and thence the gravel passes over a finer-gathering, broad undercurrent sluice; then, again, by a narrow flume, winding among the big boulders and through narrow crevices in the rocks, to a final undercurrent, where the finest material of all is collected on burlap, or sacking material.

DETAIL DESCRIPTION OF THE WORK.

It is necessary to explain the details of the work which the accompanying sketches represent, as it is more or less complicated. After all the machinery, flumes, sluices, pipes, and Giants were in place, the excavation of the pit was commenced and carried on down to bed-rock, with the aid of the Giants and elevators. When the Giant nozzles had been brought to play, the material, as the pit deepened, was forced up through the gravel elevators into the gravel sluice.

THE LUDLUM ELEVATOR.

The Ludlum gravel elevator, shown in Fig. 48, is a big steel pipe somewhat funnel shaped towards the lower portion,

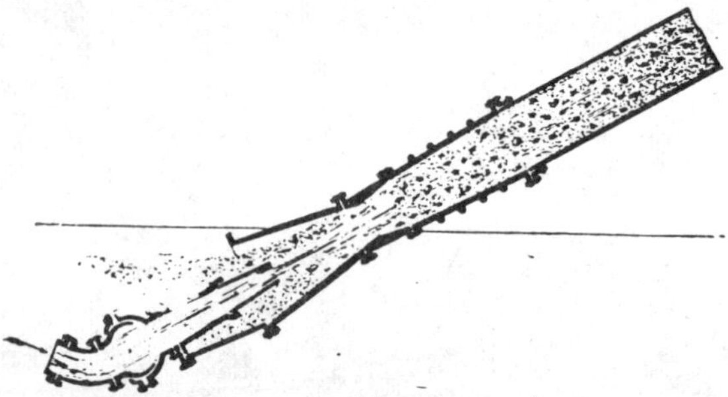

FIG. 48.

the broad end of which descends into the bottom of the pit, where both water and gravel accumulate from the work of the Giants. Directly underneath the end of the elevator, at a distance below it of 16 inches, is embedded a Giant nozzle, together with a portion of its pipe, receiving a powerful pressure from one of the main pipes on the bank. As the gravel and stones roll down they are directed by a box in upon this nozzle, the lower portion of which is enclosed in bed-rock when bed-rock is attained. The stream carries the smaller boulders and débris up the funnel of the elevator and into the flume above, where a pipe communicating with the main flume sends a flood of water into the gravel sluice to help push along the boulders and gravel that have thus come up. The other pipe that is also seen entering the end of the box of the sluice and passing down in a steep, slanting direction into the pit is a Ludlum water-lifter, sometimes called an "elevator pump." It works somewhat like the gravel elevator, a vacuum being formed in the lower portion of the pipe, drawing the water of the pit up into it. The power-pressure nozzle runs about a foot up into the pipe. The purpose of this pump is to drain the pit of the water accumulating from the Giants and in other ways. So, while the Giants tear down the banks, the elevators carry the water, gravel, and gold up into the gravel sluices. The main gravel sluice is a narrow trough or box, 208 feet long by 48 inches wide and 3 feet deep, laid down at a gentle inclination on the top of the surface of the creek bed from the lower end of the excavation. It is made of strong, inch boards and paved on the bottom with square, 8-inch blocks of pine wood set on end, so that the grain is uppermost. These block riffles are laid in rows quite close together across the bottom of the sluice, from side to side. Between each set or row of blocks is laid a narrow strip of wood, 3 inches high by ½ inch thick.

In laying block riffles, the blocks in the first row are placed closely side by side. Then the strip of wood is nailed along the lowest part of them with headless nails, not driven home, but protruding a little, so that when the next row of riffles is laid down and driven up, the protruding nails sink into the blocks and hold them fast while the strip is being laid against the lower side. The gravel, as it is being borne along in the sluice, drops its gold, which is collected in these cracks or gaps between the riffles, prepared to receive it.

On the side of this main sluice, and connected with it at the head, are two smaller side sluices, a little below it and running parallel to it. These are lined with Brussels carpet instead of riffle blocks. This carpet collects the finer gold, while the main flume usually collects the coarser material, boulders, gravel, and gold.

Towards the end of the main sluice a few of the block riffles are omitted and a grating substituted, made the full width of the sluice, with bars spaced ¾ inch apart and beveled on the bottom. This grating allows only loose stones or gravel below a certain size to pass, together with finer material, into the next sluice, called an "undercurrent." This is a broad, shallow box, similar to that shown in Fig. 36, tipped at an inclination of 6 inches in 24 feet, the latter being the length of the undercurrent, which is 12 feet wide. The bottom of this box is lined with a peculiar kind of riffle. These riffles consist of narrow slats or strips of wood, laid down on the bottom, across the width of the box, and on top of each slat is a piece of strap iron, nailed flat, whose edge overlaps the slat on both sides, but only slightly on the lower side. The water passing through these moves to and fro, like an endless pulley, from riffle to riffle, dropping its gold among them by the eddies so caused. There still remains a certain amount of very fine material, carrying even finer gold, which escapes this first undercurrent and must not be lost. So from this a narrow flume, winding through a passage in the rock, leads into a still larger, longer, and wider undercurrent, which catches the finest material of all—in this case composed largely of finely comminuted pyrites, the tailings from the mills. This long, wide undercurrent, 45 feet long by 24 feet wide, is divided into a series of compartments or boxes, set longitudinally. The divisions are formed by long boards a foot deep. At the bottom of these boards a narrow strip of wood is laid and battened down on the burlap, or sacking material, which lines the bottom of the box and receives the gold. The burlap carpets are drawn off by rollers on swivels and transported to a wooden tank, where they pass over a series of rollers, laying them conveniently open for inspection and cleaning. Every visible particle of gold is collected, and the rest drops into the water in the tank. Through the middle of this undercurrent sluice passes a small flume with perforated plates at the upper end. This flume is intended to catch and dispose of some of the coarser material that may have passed through the upper undercurrent, and what finer gold there may be in it drops through the perforated plates into the general undercurrent, the coarser rubbish being carried out to the river. On cleaning-up days, which occur at intervals, the block riffles are taken up and carefully inspected for gold. This leaves the bottom of the sluice uncovered, and on this a good deal of gravel, gold, and quicksilver has collected. This is carefully shoveled into buckets and examined, the gold laid aside, and the quicksilver amalgam containing gold placed in retorts; so also in the other undercurrents, together with the Brussels carpet and burlaps in both sluices and undercurrents.

The bed of the stream, as at present excavated by the

pit, shows a section of the placer. (Fig. 49.) The great loose rocks, by forming the so-called stone dam across the stream, produced a natural gathering place and stoppage for all the boulders and rubbish brought down by floods from above. Some of the boulders are several feet in diameter and have to be blasted before they can be removed. Mixed with these boulders are

FIG. 49.

many stumps and logs of driftwood, some of which show the marks of beavers. Half way down the section is a dark line, formed by a thin bed of peat, marking the origin of an old surface soil. Above this are belts of irregularly bedded gravel and sand, showing the action of shifting currents. Gold has been found all the way down from surface to bed-rock, the coarsest and most abundant gold being on bed-rock itself. They are obliged to wall up portions of the loose sides of the pit with cobblestones, as the jarring of passing trains is likely to shake down boulders, endangering the lives of the workmen and gradually undermining the adjacent railroad tracks.

THE END.

SHOREY PUBLICATIONS IN PRINT

1.	Glover, S.L. ORIGIN AND OCCURENCE OF GEM STONES IN WASHINGTON. 1949	1.00
2.	Mooney, Jas. Ghost Dance Religion: SMOHALLA AND HIS DOCTRINE. 1896	1.50
3.	Mooney. Ghost Dance Religion: SHAKERS OF PUGET SOUND. 1896	1.50
5.	DICTIONARY OF CHINOOK JARGON. nd	2.00
6.	ALASKA RAILROAD TIME TABLES. 1922	.50
7.	Wells, E.H. UP AND DOWN THE YUKON. 1900	1.50
8.	Williams, J.G. REPORT OF ATTORNEY GENERAL, Territory of Alaska, 1949-51.	1.50
9.	Gruening, Ernest. MESSAGE OF THE GOVERNOR OF ALASKA, 1946.	2.00
10.	Gruening. THIRD MESSAGE TO THE PEOPLE OF ALASKA, 1946.	1.50
11.	Kohlstedt, E.D. A GLIMPSE OF ALASKA. 1930	1.00
12.	Gruening. MESSAGE OF THE GOVERNOR OF ALASKA, 1945.	2.50
13.	Leehey, M.D. PUBLIC LAND POLICY OF THE U.S. IN ALASKA. 1912	1.50
14.	NOME TELEPHONE DIRECTORY. 1905	3.00
15.	ESTABLISHMENT OF MT. MCKINLEY NATIONAL PARK. 1916	1.50
16.	Hooper, Capt. C.L. CRUISE OF THE U.S. REVENUE STEAMER CORWIN IN ARCTIC OCEAN, November 1, 1880.	7.50
17.	Gruening. MESSAGE TO THE PEOPLE OF ALASKA. 1945	2.00
18.	Gruening. MESSAGES OF THE GOVERNOR OF ALASKA. 1949	2.50
19.	Davis, Geo. T. METLAKAHTLA. 1904	3.50
20.	WORK OF THE BUREAU OF EDUCATION FOR NATIVES OF ALASKA. 1918	5.00
21.	MacDowell, L.W. ALASKA INDIAN BASKETRY.	1.00
22.	Bells, Rev. Myron. THE TWANA, CHEMAKUM & KLALLAM INDIANS OF WASHINGTON TERRITORY. 1887	4.00
23.	Grinnell, Joseph. GOLD HUNTING IN ALASKA.	4.00
24.	Swan, James G. INDIANS OF CAPE FLATTERY.	7.50
25.	CUSTER'S LAST BATTLE. 1892	2.00
26.	Gruening, Sen. INDEPENDENCE DAY ADDRESS. 1959	1.75
27.	Cadell, H.M. THE KLONDIKE AND YUKON GOLDFIELD IN 1913. 1914	2.50
28.	Shiels, Archie W. EARLY VOYAGES OF THE PACIFIC. 1930	3.00
29.	Fickett, E.D. METEOROLOGY. From Explorations in Alaska. 1900	
30.	Richardson, W.P. & others. YUKON RIVER EXPLORING EXPEDITION. Expl. in Alaska. 1900	2.00
31.	Ray, P.H., W.P. Richardson. RELIEF OF THE DESTITUTE IN THE GOLDFIELDS. 1900	1.75
32.	Shiels, Archie W. LITTLE JOURNEYS INTO THE HISTORY OF RUSSIAN AMERICA AND PURCHASE OF ALASKA. 1949	7.50
33.	Shiels. STORY OF TWO DREAMS. 1957	4.00
34.	Swan, J.G. HAIDAH INDIANS OF QUEEN CHARLOTTE'S ISLAND, B.C. 1874	5.00
35.	Butler, Gen. B.V. & the Marquis of Lorne. THE BERING SEA CONTROVERSY. 1892	1.50
36.	Crawford, Lewis F. THE MEDORA-DEADWOOD STAGE LINE. 1925	1.50
37.	Anderson, Eskil. ASBESTOS AND JADE OCCURENCES IN KOBUK RIVER REGION, Alaska. 1945	2.00
38.	Stewart, B.D. PROSPECTING IN ALASKA. 1949	1.50
39.	Geoghegan, Richard H. THE ALEUT LANGUAGE.	5.00
40.	SEATTLE'S FIRST BUSINESS DIRECTORY. 1876	7.50
41.	THE WASHBURN YELLOWSTONE EXPEDITION. 1871	1.25
42.	Subreports from EXPLORATIONS IN ALASKA. 7 reports: The Tanana, Chickaloon, Sushitna, etc. 1900	2.00
43.	Glenn, Capt. E.F. TANANA RIVER EXPLORING EXPEDITION. 1900	2.50
44.	Glenn. COOK'S INLET EXPLORING EXPEDITION. 1900	1.75
45.	WILKESON'S NOTES ON PUGET SOUND (1870?)	2.50
46.	(Perry). JOURNAL OF A VOYAGE TO THE ARCTIC REGIONS IN H.M.S. ALEXANDER, 1818. (1819?)	6.00
47.	BIENNIAL MESSAGE OF WM. M. BUNN, Governor of Idaho. 1884	2.00
48.	MACKENZIE'S ROCK. Exploration of Sir Alexander Mackenzie. 1905	2.50
49.	Williams, L.R. OUR PACIFIC COUNTY (Wash)	5.00
50.	AARON LADNER LINDSLEY..Founder of Alaska Missions.	1.00
51.	Evans, Elwood. PUGET SOUND: Its Past, Present and Future. 1869	2.50
52.	Campbell, Robt. TWO JOURNALS, 1808-1853.	15.00
53.	Bernhardi, Madame Charlotte. MEMOIR OF THE CELEBRATED ADM. JOHN de KRUSENSTERN. 1856	5.00
54.	McWhorter, Lucullus V. CRIME AGAINST THE YAKIMAS. 1931	4.50
55.	Lupton, Chas T. OIL AND GAS IN THE OLYMPIC PENINSULA. 1913	4.00
56.	Seward, Wm. H. THE ADMISSION OF KANSAS. A Speech. 1860	1.50
57.	Stevens, Issac I. PROCLAMATION OF MARTIAL LAW. 1856	.50
58.	THE GLACIER - Tlinkit Training Academy, Vol. II. 1887	.75
59.	Hills, Rev. E.P. REV. AARON L. LINDSLEY.	.50
60.	SWANTON'S HANDBOOK OF NORTH AMERICAN INDIANS, Pt. C: Indians of Alaska & Canada. 1952	5.00
61.	Ward, D.B. ACROSS THE PLAINS IN 1853. 1911	3.50
62.	MacDowell, Lloyd W. ALASKA TOTEM POLES.	1.25
63.	MacDowell. A TRIP TO WONDERFUL ALASKA.	2.00
64.	MacDowell. ALASKA GLACIERS & ICE FIELDS.	1.25
65.	Shaw, Geo. C. THE CHINOOK JARGON AND HOW TO USE IT. 1909	4.00

66. PUGET SOUND COOPERATIVE COLONY. 1890's 2.00
67. THE HEROES OF BATTLE ROCK, or The Miners' Reward. 1904 1.50
68. Lampman, Ben Hur. CENTRALIA TRAGEDY AND TRIAL. 1920 4.00
69. SEAL FISHERY IN THE NORTH PACIFIC. (Correspondence with Russia.) 1895 7.50
70. Coolican, J.C. (ed.) PORT ANGELES. 1898 4.00
71. Jordan, D.S. TROUT AND SALMON OF THE PACIFIC COAST. 1906 1.25
72. COOPERATIVE PLAN FOR SECURING HOMES AT PORT ANGELES, Wash. 1893 1.00
73. Hidden, Maria L.T. OREGON PIONEERS. 1910 1.00
74. Stevens, Hazard. FIRST ASCENT OF TAKHOMA. Ext., 1876 1.50
75. DEMOCRATIC AND REPUBLICAN Ticket For Washington State and King County. 1889 .50
76. THIRD ANNUAL COMMENCEMENT OF SEATTLE HIGH SCHOOL. 1888 .35
77. ASSASSINATION OF PRESIDENT GARFIELD. Newspaper Extra. 1881? .50
78. PUGET SOUND CATECHISM. 1889 1.00
79. SEATTLE, KING COUNTY, WASHINGTON TERRITORY. Statistics & descriptive report. 1884 1.50
80. THE GREAT SEATTLE FIRE OF June 6, 1889. 2.50
81. USGS Bull. 442-D. MINING IN THE CHITINA DIST., etc. 1910 2.50
82. Swanton, John R. INDIAN TRIBES OF THE PACIFIC NORTHWEST. Ext., 1952 6.00
83. Sayre, Alex N. PUGET SOUND. Poem 1883 1.50
84. Bell, W.S. OLD FORT BENTON. 1909 2.50
85. Meany, Edmund S. INDIAN GEOGRAPHIC NAMES OF WASHINGTON. 1908 1.00
86. Swanton, J.R. INDIAN TRIBES OF AMERICAN SOUTHWEST. 1952 6.00
87. Swanton. INDIANS OF MEXICO, CENTRAL AMERICA AND WEST INDIES. 1952 2.50
88. USGS Bull. 836. SELECTED LIST OF USGS PUBLICATIONS ON ALASKA. 1930 1.00
89. Lewis, Wm. S. EARLY DAYS IN THE BIG BEND COUNTRY. 1926 3.50
90. Stoney, Geo. M. EXPLORATIONS IN ALASKA. 1899, Ext. from US Naval Inst. 6.00
91. Lindsley, A.L. SKETCHES OF AN EXCURSION TO SOUTHERN ALASKA. 1881 5.00
92. Collins, H.B. ARCHEOLOGY OF THE BERING SEA REGION. Ext., 1933 1.50
93. Ray and Murdoch. A VOCABULARY OF THE ESKIMOS OF POINT BARROW. 1885 1.25
94. Andrews, C.L. STORY OF SITKA. 1922 5.00
95. Matthews, Washington. NAVAJO WEAVERS. 1884 2.50
96. Dawson, Geo. M. THE HAIDAS. 1882 1.00
97. McClintock, Walter. FOUR DAYS IN A MEDICINE LODGE. 1900 1.00
98. Barbeau, Marius. MODERN GROWTH OF THE TOTEM POLES ON THE NORTHWEST COAST. 1939 1.00
99. SCRIPTURE SELECTIONS AND HYMNS IN HIDATSA or GROS VENTURE LANGUAGE. 1906 1.50
00. Baker, Marcus. GEOGRAPHIC DICTIONARY OF ALASKA. 1906 3.00
01. Lee, Charles A. ALEUTIAN INDIAN AND ENGLISH DICTIONARY. 1896 1.00
02. Shaw, Geo. C. VANCOUVER'S DISCOVERY OF PUGET SOUND IN 1792. 1933 1.00
03. Seward, Wm. H. SPEECHES in Alaska, Vancouver, and Oregon, 1869. 2.50
04. Rees, J.E. IDAHO: Chronology, Nomenclature, Biblio. 1918 5.00
05. Smith, W.C. THE EVERETT MASSACRE. 1916 8.00
06. PLACER MINING. 1897 4.50
07. Ross, Clyde P. THE VALDEZ CREEK MINING DISTRICT IN ALASKA. 1933 5.00
08. Stevens, Isaac I. TREATY BETWEEN THE U.S. AND THE DWAMISH, SUQUAMISH, & OTHER ALLIED AND SUBORDINATE TRIBES OF INDIANS IN WASHINGTON TERRITORY. 1855 2.00
09. Stevens, I.I. TREATY BETWEEN THE U.S. & THE NISQUALLY & OTHER BANDS OF INDIANS. 1855 2.00
10. Stevens. TREATY BETWEEN THE U.S. & THE YAKIMA NATION OF INDIANS. 1855 2.00
11. Stevens. TREATY BETWEEN THE U.S. OF A. & THE MAKAH TRIBE OF INDIANS. 1855 2.00
12. Stevens. TREATY BETWEEN THE U.S. AND THE INDIANS OF THE WILLAMETTE VALLEY. 1855 2.00
13. Densmore, F. STUDY OF INDIAN MUSIC. 1941 1.50
14. Harriman, Job. THE CLASS WAR IN IDAHO. 3.00
15. Ranck, G. PICTURES FROM NORTHWEST HISTORY. 2.00
16. Whitaker, R. OVERLAND TO OREGON. 1906 4.00
17. Krieger, H. INDIAN VILLAGES OF S.E. ALASKA. 2.50
18. Gatschet, A. THE KLAMATH INDIANS OF SOUTHWEST OREGON. 1890 4.50
19. Crane, W. TOTEM TALES. 1952 4.00
20. Bate, J. THE SECOND BOOKE TEACHING MOST PLAINLY AND COMPOSING OF ALL MANNER OF FIREWORKS FOR TRYUMPH AND RECREATION. 1635 3.00
21. James, G. PRACTICAL BASKET MAKING. 1917 5.00
22. Whorf, B. MAYA HIEROGLYPHS. 1941 1.25
23. Dall, Wm. MASKS, LABRETS, AND CERTAIN ABORIGINAL CUSTOMS. 1884 8.50
24. McCurdy, J. CAPE FLATTERY AND ITS LIGHT. 1.00
25. Wilhelm, Homer. THE SAN JUAN ISLANDS. 1901 5.00
26. Gaerity, Jack. BREAD AND ROSES FROM STONE. 4.00
27. Smith, W.C. WAS IT MURDER? (Centralia) 1922 3.50
28. HISTORY OF OWYHEE COUNTY. (Idaho) 8.50
29. Evans, Elwood. WASHINGTON TERRITORY. 1877 3.00
30. Delaney, Matilda Sagar. THE WHITMAN MASSACRE. 1920 3.00
31. Hubback, T.W. TEN THOUSAND MILES TO ALASKA FOR MOOSE AND SHEEP. 1921 4.00
32. Wilson, Katherine. COPPER-TINTS. 1923 3.00
33. Coleman, Edmund T. FIRST ASCENT OF MT. BAKER. Ext. from Harper's Monthly. 1869 2.50
34. Hodge, L.K. (ed.) MINING IN THE PACIFIC NORTHWEST. 1897 30.00
34a. Hodge. MINING IN WESTERN WASHINGTON. Ext from #134 10.00

134b. Hodge. *MINING IN CENTRAL & EASTERN WASHINGTON.* Ext. from #134 10.00
134c. Hodge. *MINING IN SOUTHERN BRITISH COLUMBIA.* Ext. from #134 10.00
135. Stallard, Bruce. *ARCHEOLOGY IN WASHINGTON.* 1958 3.00
136. Lockley, Fred. *ALASKA'S FIRST FREE MAIL DELIVERY IN 1900.* 1.25
137. Strange. *JAMES STRANGE'S JOURNAL AND NARRATIVE OF THE COMMERCIAL EXPEDITION FROM BOMBAY TO THE NORTHWEST COAST OF AMERICA.* 1928 5.00
138. Hathaway, Ella C. *BATTLE OF THE BIG HOLE.* 2.00
139. Haller, Granville. *SAN JUAN & SECESSION.* 1.50
140. *THE KLONDIKE NEWS.* Vol. 1 #1, Dawson Newspaper. 1898 (Alaska's rarest newspaper) 10.00
141. *JOURNAL OF MEDOREM CRAWFORD.* 1897 2.00
142. *OCOSTA! The Ocean Terminus of the Northern Pacific R.R. and Coast City of Washington.* 2.50
143. Sutherland, T.A. *HOWARD'S CAMPAIGN AGAINST THE NEZ PERCE INDIANS, 1877.* 1878 3.50
144. Matthews, Mathew. *THE CATLIN COLLECTION OF INDIAN PAINTINGS.* 1890 3.00
145. Lockley, Fred. *TO OREGON BY OX TEAM IN '47.* 1.25
146. Lockley. *VIGILANTE DAYS IN VIRGINIA CITY.* 1.50
147. Judson, Katherine. *MYTHS AND LEGENDS OF THE PACIFIC NORTHWEST.* 1910 7.50
148. Whiting, Dr. F.B. *GRIT, GRIEF AND GOLD.* 10.00
149. Amundsen, Capt. Roald. *TO THE NORTH MAGNETIC POLE AND THROUGH THE NORTHWEST PASSAGE.* 3.00
150. Butte Businessmen's Assn. *BUTTE, MONTANA.* 2.00
151. Immigration Aid Society of Washington Terr. *NORTHWESTERN WASHINGTON.* 1880 3.50
152. Denig, Edwin T. *INDIAN TRIBES OF THE UPPER MISSOURI.* 1930 10.00
153. Buckley, Rev. J.M. *TWO WEEKS IN THE YOSEMITE AND VICINITY.* 1888 2.50
154. *SPEECH OF THE HON. R.C. WINTHROP OF MASS. ON THE PRESIDENT'S MESSAGE (on admission of California)* 1850 2.00
155. Marshall, Martha. *A PRONOUNCING DICTIONARY OF CALIFORNIA NAMES IN ENGLISH AND SPANISH.* 2.50
156. Mercer, A.S. *BIG HORN COUNTY, WYOMING.* 7.50
157. Moorehead, Warren K. *PREHISTORIC RELICS.* 7.50
158. Merrill, Geo. P. *NOTES ON THE GEOLOGY AND NATURAL HISTORY OF THE PENINSULA OF LOWER CALIFORNIA.* 1895 2.00
159. Georgeson, C.C. *REINDEER & CARIBOU.* 1904 1.50
160. Hewett, Edgar L. *A GENERAL VIEW OF THE ARCHEOLOGY OF THE PUEBLO REGION.* 1904 2.00
161. Bennett, Wm. P. *THE FIRST BABY IN CAMP.* 3.50
162. Coffman, Noah B. *OLD LEWIS COUNTY, Oregon Territory.* 1926 3.50
163. Bebbe, Mrs. Iola. *THE TRUE LIFE STORY OF SWIFTWATER BILL GATES.* 1908 7.50
164. Costello. *THE SIWASH - THEIR LIFE, LEGENDS AND TALES.* 1895 10.00
165. Murie, Olaus J. *ALASKA-YUKON CARIBOU.* 10.00
166. Meeker, Ezra. *STORY OF THE LOST TRAIL TO OREGON, NO. 2.* 1916 2.50
167. Hanna, Rev. J.A. *DR. WHITMAN AND HIS RIDE TO SAVE OREGON.* 1903 1.25
168. Hadwen, Seymore & Laurence J. Palmer. *REINDEER IN ALASKA.* 1922 5.00
169. Geer, T.T. *THE ROMANCE OF ASTORIA.* 1911 1.75
170. Williams, Lewis. *CHINOOK BY THE SEA.* 7.50
171. *SAN FERNANDO VALLEY.* 1938 2.00
172. *PASADENA.* 1938 2.00
173. Ingalls, Maj. J.W. *HISTORY OF WASHOE COUNTY, NEVADA.* 1913 5.00
174. Reid, J.T. & J.R. Hunter. *HUMBOLT COUNTY.* 1913 4.50
175. Aston. *ESMERALDA COUNTY.* 1913 4.50
176. London, Jack. *THE GOLD HUNTERS OF THE NORTH.* Ext., 1903. 1.25
177. Wartman-Arland, Flora E. *THE STORY OF MONTESANO.* 1933 4.00
178. London. *THE ECONOMICS OF THE KLONDIKE.* 1.50
179. Bishop, Robert Sr. *LAND IN THE SKY TOTEM.* 1.50
180. Webb, John S. & Ed S. Curtis. *THE RIVER TRIP TO THE KLONDIKE AND THE RUSH TO THE KLONDIKE OVER THE MOUNTAIN PASS.* 1898 2.50
181. Librn. of Bellingham Public Libr. *HISTORY OF BELLINGHAM.* 1926 6.00
182. *VOYAGE OF ALEXANDER MACKENZIE.* 2.50
183. Chaplin, Ralph. *CENTRALIA CONSPIRACY.* 4.50
184. Haswell, Robert. *ROBERT HASWELL'S JOURNALS.* 1788-89. 4.00
185. Minto, John. *RHYME OF EARLY LIFE IN OREGON.* (1915?) 4.00
186. Collins, Henry B., A.H. Clarke & E.H. Walker. *THE ALEUTIAN ISLANDS: Their People & Natural History.* 1945 7.50
187. Simpson, Brevet Brig. Gen. J.H. *CORONADO'S MARCH IN SEARCH OF THE "SEVEN CITIES OF CIBOLA."* 1871 2.50
188. Riddle, Geo. W. *EARLY DAYS IN OREGON.* 5.00
189. *VILHJALMUR STEFANSON.* 1925 2.50
190. Lockley, Fred. *ACROSS THE PLAINS BY PRAIRIE SCHOONER.* 1.50
191. MacArthur, Walter. *LAST DAYS OF SAIL ON THE WEST COAST.* 1929 7.50
192. Stuck, Hudson. *T E ALASKAN MISSIONS OF THE EPISCOPAL CHURCH.* 1920 10.00
193. Steffa, Don. *TALES OF NOTED FRONTIER CHARACTERS, SOAPY SMITH.* 1908 1.75
194. Meeker, E. *WASGUBGTIB TERRUTIRT,* k870 3.00
195. Denny, Arthur A. *PIONEER DAYS ON PUGET SOUND.* 1888 5.00
196. Rowan, James. *THE I.W.W. IN THE LUMBER INDUSTRY.* 4.00
199. Sayre, J. Willis. *THE EARLY WATERFRONT OF SEATTLE,* 1937. 2.50

204. Pelly, T.M. DR. MINOR - A SKEtch of His Background and Life. 1933 — 7.50
209. Flandrau, Grace. FRONTIER DAYS ALONG THE UPPER MISSOURI. — 3.00
210. Flandrau. THE LEWIS AND CLARK EXPEDITION. — 5.00
211. Flandrau. A GLANCE AT THE LEWIS AND CLARK EXPEDITION. — 3.00
213. Flandrau. THE VERENDRYE OVERLAND QUEST OF THE PACIFIC. — 3.50
216. Harper, Frank B. FORT UNION AND ITS NEIGHBORS ON THE UPPER MISSOURI. — 2.50
220. Burdick, Usher L. MARQUIS de MORES AT WAR IN THE BADLANDS. 1929 — 2.50
227. General Strike Committee. THE SEATTLE GENERAL STRIKE. 1919 — 5.00
231. Sayre, J.W. THE ROMANCE OF SECOND AVENUE. — 1.50
235. DEDICATION AND OPENING OF THE NEW CASCADE TUNNEL. 1929 — 2.00
239. Van Olinda, O.S. HISTORY OF VASHON - MAURY ISLAND. 1935 — 6.00
241. Hanford, C.H. SAN JUAN DISPUTE. 1900 — 2.00
246. Walgamott, C.S. REMINISCENCES OF EARLY DAYS. 1926 — 7.50
250. Wardner, Jim. JIM WARDNER, OF WARDNER IDAHO. 1900 — 8.00
251. Brown, Col. W.C. THE SHEEPEATER CAMPAIGN, Idaho - 1879. 1926 — 2.50
259. Hornaday, Wm. T. EXTERMINATION OF THE AMERICAN BISON. 1887? — 10.00
264. Slauson, Morda C. ONE HUNDRED YEARS ON THE CEDAR. — 10.00
265. Jones, S.C. & M.F. Casady. FROM CABIN TO CUPOLA. County Courthouses in Washington. — 10.00
267. Andrews, Clarence L. WRANGELL AND THE GOLD OF THE CASSIAR. 1937 — 3.50
268. Andrews, Clarence L. THE PIONEERS AND THE NUGGETS OF VERSE. 1937 — 3.00
280. PORT TOWNSEND. 1890 — 5.00
281. Gilbert, Kenneth. ALASKAN POKER STORIES. 1958 — 2.50
282. Moore, James Bernard. SKAGWAY IN DAYS PRIMEVAL. — 4.00
283. Chase, Cora G. UNTO THE LEAST. 1972 — 6.00
284. Blankenship, Geo. E. EARLY HISTORY OF THURSTON COUNTY. 1914 — 20.00
285. Fish, Harriet U. PAST AT PRESENT. 1967 — 12.50
286. Allen, Edward W. THE ROLLICKING PACIFIC; A Selection of Poems. 1972 — 3.00
287. Allen. DANCING TALES. 1951 — 4.00

SJU 1. Farlow, Dr. W.G. SOME EDIBLE AND POISONOUS FUNGI. 1897 — 2.50
SJU 2. Gurdji, V. ORIENTAL RUG WEAVING. 1901 — 5.00
SJU 3. D'Avenes, E., Prisse & Dr. J.C. Ewart. EGYPTIAN AND ARABIAN HORSES AND ORIGIN OF HORSES AND PONIES. 1904 — 2.50
SJU 4. Freshfield, Douglas W. ON MOUNTAINS AND MANKIND. 1904 — 1.50
SJU 5. Davenport, Cyril. CAMEOS. 1904 — 1.50
SJU 6. Maire, Albert. MATERIALS USED TO WRITE UPON BEFORE THE INVENTION OF PRINTING. — 1.50
SJU 7. Liberty, Arthur L. PEWTER AND THE REVIVAL OF ITS USE. 1904 — 1.50
SJU 8. Hammell, Wm. PINE NEEDLE BASKETRY. — 1.50
SJU 9. Frey, Clark, Vietch. HOME TANNING. 1936 — 1.50
SJU 10. Browning, Frank. STEAM PLANT ERRORS — 2.00
SJU 18. Burdick, Arthur J. THE PROSPECTOR'S MANUAL. 1905 — 7.50
SJU 19. WOODS DUAL POWER. (1902?) — 1.50
SJU 20. Hayward, Charles B. DIRIGIBLE BALLOONS. 1921 — 5.00
SJU 21. Osborn, Henry F. THE ELEPHANTS AND MASTEDONS ARRIVE IN AMERICA — 1.50
SJU 22. Todd, Mattie Phipps. HAND LOOM WEAVING. 1902 — 7.00
SJU 23. Wilbur, C. Martin. HISTORY OF THE CROSSBOW. 1936 — 1.50
SJU 24. Alexander, A.S. HORSE SECRETS. 1913 — 2.50
SJU 26. Clute, Willard Nelson. THE FERN - COLLECTOR'S GUIDE. 1901 — 3.00
SJU 27. Meeker, E. HOP CULTURE IN THE U.S. 1883 — 10.00

SJI 1. Krause, F. SLING CONTRIVANCES. 1904 — 2.00
SJI 2. Murdoch, John. A STUDY OF THE ESKIMO BOWS IN THE U.S. NATIONAL MUSEUM. 1884 — 1.50
SJI 3. Mason, Otis T. THROWING STICKS IN THE NATIONAL MUSEUM. 1884. — 2.50
SJI 4. CHIEF JOSEPH'S OWN STORY. 1879 — 2.50
SJI 5. Mason, Otis. BASKET WORK OF THE ABORIGINES. 1884 — 5.00
SJI 6. Llwd, Rev. Dr. J.P.D. THE MESSAGE OF AN INDIAN RELIC: Seattle's Totem Pole. — 2.00
SJI 7. Mason, O. TRAPS OF THE AMERICAN INDIAN. — 1.25
SJI 8. Meeker, Louis L. OGALALA GAMES. 1901 — 1.50
SJI 9. Collins, Henry B. PREHISTORIC ART OF THE ALASKAN ESKIMO. 1929 — 3.50
SJI 10. Eells, Rev. M. JUSTICE TO THE INDIAN. — 1.00
SJI 11. Willoughby, C. INDIANS OF THE QUINAIELT AGENCY, Washington Territory. 1886 — 1.50
SJI 12. Leon & Holmes. STUDIES ON THE ARCHAEOLOGY OF MICHOACAN MEXICO... 1886 — 2.00
SJI 13. Beckwith, Paul. NOTES ON CUSTOMS OF THE DAKOTAHS. 1886 — 1.25
SJI 14. Eells, Rev. M. THE STONE AGE. 1886 — 1.50
SJI 15. Allen, Lt. Henry T. ATNATANAS NATIVES OF COPPER RIVER, ALASKA. 1886 — 1.00
SJI 16. Yates, Dr. L.G. CHARM STONES. 1886 — 1.50
SJI 17. Boas, Fran . THE CENTRAL ESKIMO. 1884 — 12.50
SJI 18. Gibbs, Geo. & others. TRIBES OF THE EXTREME NORTHWEST, ALASKA, THE ALEUTIANS AND ADJACENT TERRITORIES. Ext. 1877 — 15.00
SJI 19. Gibbs & others. LANGUAGES OF THE TRIBES OF THE EXTREME NORTHWEST, ALASKA, THE ALEUTIANS AND ADJACENT TERRITORY. — 6.00

SJI 20.	Gibbs, Geo., Dr. Wm. F. Tolmie, & Father G. Mengarini. *TRIBES OF WESTERN WASHINGTON AND NORTHWESTERN OREGON.*	20.00
SJI 21.	Gibbs, Geo & others. *COMPARATIVE VOCABULARIES OF THE TRIBES OF WESTERN WASHINGTON AND NORTHWESTERN OREGON.* part of SJI 20.	5.00
SJI 22.	Gibbs, Geo. *A DICTIONARY OF THE NISQUALLY INDIAN LANGUAGE.* part of SJI 20.	10.00
SJI 23.	Boas, Franz. *INTRODUCTION TO HANDBOOK OF AMERICAN INDIAN LANGUAGES.* Ext. 1911	5.00
SJI 24.	Swanton, John R. *THE TLINGIT INDIAN LANGUAGE.* From Hndbk. of Amer. Indian Lang.	3.00
SJI 25.	Swanton. *THE HAIDA INDIAN LANGUAGE.*	5.00
SJI 26.	Boas. *THE TSIMSHIAN INDIAN LANGUAGE.*	8.00
SJI 27.	Boas. *THE KWAKIUTL INDIAN LANGUAGE.*	7.50
SJI 28.	Boas. *THE CHINOOK INDIAN LANGUAGE.*	5.00
SJI 29.	Boas & Swanton. *THE SIOUAN, Dakota INDIAN LANGUAGE.*	6.00
SJI 30.	Talbitzer, Wm. *THE ESKIMO LANGUAGE.*	6.50
SJI 31.	Ewers, John C. *EARLY WHITE INFLUENCE UPON PLAINS INDIAN PAINTING.* 1957	2.00
SJI 32.	Fenton, Wm. N. *CONTRACTS BETWEEN IROQUOIS HERBALISM AND COLONIAL MEDICINE.* 1941	2.00
SJI 38.	Remington, Fred. *ARTIST WANDERINGS AMONG THE CHEYENNES.* 1889	1.50
SJI 41.	Bagley, Clarence. *INDIAN MYTHS OF THE NORTHWEST.* 1930	8.50
SJI 43.	Mason, O. *ABORIGINAL SKIN DRESSING.*	5.00
SJI 46.	Thorne, J. Frederic. *IN THE TIME THAT WAS.* 1909	2.50
SJI 49.	DeSmet, Rev. P.J. *NEW INDIAN SKETCHES.*	8.50
SJI 51.	Mason, O. *MAN'S KNIFE AMONG THE NORTH AMERICAN INDIANS.* 1899	2.00
SJI 53.	Holmes, William H. *POTTERY OF THE ANCIENT PUEBLOS.* 1886	6.00
SJI 55.	Dorsey, James O. *A STUDY OF SIOUAN CULTS* 1894	10.00
SJI 57.	Mason, O. *POINTED BARK CANOES OF THE KUTENAI AND AMUR.* 1901	2.00
SJI 58.	Mason, O. *ABORIGINAL AMERICAN HARPOONS.* 1902	7.50
SJI 59.	Hough, Walter. *THE LAMP OF THE ESKIMO.* 1898	5.00
SJI 61.	Drucker, Philip. *ARCHAEOLOGICAL SURVEY ON THE NORTHERN NORTHWEST COAST.* 1943.	7.00
SJI 62.	*INDIANS IN WASHINGTON.*	3.00

NOTE TO STUDENTS AND COLLECTORS

Many scarce items turn up in antiquarian book stores only once. And while such works are in great demand, the high prices usually commanded by the older, scarcer items put them beyond the reach of the average student and the small collector. To meet this need we are bringing back into print a diversity of Pacific Northwest and Alaskan historical material which we are selling at moderate prices. In their original form these bring prices ranging from $3.00 to $100.00; some even higher.

We limit most reproductions from 25 to 100 copies and reprint as the demand warrants. The books have heavy paper covers. They show both original and reprinting dates. Regular library and dealer discounts granted. For detailed descriptions of these titles send for Shorey's Catalog of Publications #3. It also lists future Shorey Publications.